Ultralight Aviation Series / No. 5

Ultralight Technique

How To Fly And Navigate Ultralight Air Vehicles

Michael A. Markowski

Ultralight Technique —
How to Fly and Navigate Ultralight Air Vehicles
by Michael A. Markowski

Copyright © 1983 by Michael A. Markowski

FIRST EDITION
First Printing — July 1983

Published by:
Ultralight Publications, Inc.
Post Office Box 234
Hummelstown, PA 17036

All rights reserved, including the right to reproduce this book or portions thereof in any form or by any means, electronic or mechanical, including photocopying, recording, or by an information storage or retrieval system without written permission from the author, except for the inclusion of brief quotations in a review. All rights are also reserved for translation into foreign languages. No liability is assumed with respect to use of the information herein, nor is endorsement of any product implied or intended.

Books by the Author
THE HANG GLIDER'S BIBLE, The Complete Pilot's and Builder's Guide.
THE ENCYCLOPEDIA OF HOMEBUILT AIRCRAFT, including Powered Hang Gliders.
ULTRALIGHT AIRCRAFT, The Basic Handbook of Ultralight Aviation
ULTRALIGHT FLIGHT, The Pilot's Handbook of Ultralight Knowledge
ULTRALIGHT TECHNIQUE, How to Fly and Navigate Ultralight Air Vehicles

Library of Congress Cataloging in Publication Data
Markowski, Michael A.
Ultralight Technique, How to Fly and Navigate Ultralight Air Vehicles
(Ultralight Aviation Series, No. 5)
1. Ultralight Aircraft — Piloting
I. Title. II. Series.

| TL685.1.M37 | 1983 | 629.132'524 | 83-50057 |

ISBN 0-938716-12-3 (pbk.)

On The Cover
The Kolb "Ultrastar" incorporates conventional stick and rudder three-axis controls, including full span ailerons. It features a welded steel tubing fuselage and strut-braced wings that fold for a ten minute set-up. Power is supplied by the 35 hp Cuyuna ULII-02. Courtesy Kolb Company, Inc., R.D. 3, Box 38, Phoenixville, PA 19460.

To My Dad

With deep appreciation for your introducing me to the power of positive thinking, and teaching me the principles of success.

Our knowledge is a torch of smoky pine
That lights the pathway but one step ahead
Across a void of mystery and dread.
      George Santayana

WARNING — A WORD OF CAUTION

Flight, in and of itself, is not necessarily dangerous, however it is most unforgiving of errors, sloppiness and misjudgment on the part of both the designer and pilot. Whenever a man flies, he accepts the risk that he may be injured or even killed. It is each individual's decision to either accept or reject this risk in light of its potential hazards, challenges and rewards. Flying can be and is done safely every day of the year by paying strict attention to the details.

This book is not intended as a do-it-yourself guide, but merely as a source of information to be used as a reference. If there is anything you don't understand, don't hesitate to ask your flight instructor. It is further recommended that you take flight training in a two-seater ultralight and ground school before you attempt flight on your own. Ultralights are real airplanes, not toys, and they must be treated with respect.

About the Author

Mike Markowski's life literally revolves around flight. He is a graduate aeronautical engineer (specialized in low speed aerodynamics) and FAA licensed private pilot who, since 1971 has devoted his life to the development of ultralight aviation. He was instrumental in initiating hang gliding in the eastern regions of the United States, and began his writing career as editor of the original *Skysurfer* magazine. Prior to that, he was employed as an advanced research engineer for Douglas and Sikorsky aircraft companies, working in the areas of advanced designs and new concepts. Since then, he has built and flown many ultralights of his own design, founded two manufacturing firms, ran a flight school, taught ultralight flight theory at the university level (M.I.T. and others), and is considered an authority on ultralight aviation.

He has lectured to numerous organizations and groups, including the International Symposium on the Technology and Science of Low Speed and Motorless Flight held at M.I.T. He was also responsible for the Forum on Foot-Launched Flight held at the Annual Convention and Fly-In of the Experimental Aircraft Association during the mid-seventies. He has made "star" appearances on national television shows, is sought after for speaking engagements by various groups, and he is often quoted in books, magazines, and newspapers around the world.

Mike is a widely read author, having five books (The Hang Glider's Bible, The Encyclopedia of Homebuilt Aircraft, Ultralight Aircraft, Ultralight Flight and Ultralight Technique) to his credit—classics in their fields—as well as numerous magazine articles.

In addition to writing and flying, Mike is an aeronautical consultant to the ultralight industry for engineering design, as well as marketing, advertising, and public relations. He is also listed as the nation's only Technical Expert in Ultralight Aviation by Attorneys in the Products Liability and Transportation Legal Directory and, as such, is active in legal cases. He holds memberships in the Experimental Aircraft Association, Soaring Society of America, United States Hang Gliding Association, the Aircraft Owners and Pilots Association, and the National Association of Sport Aircraft Designers.

It is hoped that this book will foster safety through education as ultralight aviation develops and grows into recreational flying of the future.

Acknowledgement

Grateful thanks are extended to the manufacturers of ultralight aircraft and accessories for their cooperation in providing material for inclusion in this book. I would especially like to thank Kerry Cartier for his piece on wind chill factors and hypothermia, Sealord for their seaplane operations guidelines, and Ralph Taggart for his article on snow ski operations. I must not forget Vic Powell and John Ballantyne of the AOPA and ASF respectively, for their fine accomplishments in the way of ultralight operations and safety. Then too, Ken Peppard and Gary Perkins of the FAA were most helpful in the rules and regulations department. And finally, I'd like to thank John Robbins of Lightwing Insurance for his contributions to this work.

Contents

PROLOGUE	9
1 • The Principles of Ultralight Flight	15
2 • Preflight Inspection	25
3 • Starting and Taxiing	29
4 • Crow Hops	39
5 • Takeoff and Climbout	45
6 • Bank and Turn	51
7 • Mushes, Stalls and Spins	65
8 • Landing and the Traffic Pattern	71
9 • Sideslipping and Other Ways to Lose Altitude	79
10 • Crosswind Operations	83
11 • Instrumentation for Ultralights	87
12 • Density Altitude and the Koch Chart	101
13 • The Speeds to Fly an Ultralight	105
14 • Seaplane Operations	109
15 • Snow Ski Operations	119
16 • Basic Navigational Methods and Procedures	129

17 • Wind Chill Factors and Hypothermia	145
18 • Ultralight Equipment — Glossary and Function	151

APPENDICIES

A-1 • The Ultralight Solution — Separating Aircraft from Air Vehicles	167
A-2 • The Ultralight Rules — FAR Part 103	171
A-3 • Clarification of FAR 103	191
A-4 • Operation of Ultralights at Existing Airports	197
A-5 • ASF Ultralight Safety Tips	205
A-6 • ASF Ultralight Incident/Accident Reporting Form	209
A-7 • Ultralight Vehicle Airman Certification Program	213
A-8 • Ultralight Vehicle Registration Program	219
A-9 • Suggested Standards for Ultralight Flight Parks	223
A-10 • AOPA Ultralight Division Competition Rules	227
A-11 • The Powered Ultralight Manufacturers Association	241
A-12 • Canadian Ultralight Pilot License Requirements	243
A-13 • Ultralight Organizations and Publications	249
A-14 • Insurance Exposures and Definitions for Ultralight Pilots and Organizations	251

Prologue

"With each advent of spring, when the air is alive with innumerable happy creatures; when the storks on their arrival at their old northern resorts fold up their imposing flying apparatus which has carried them thousands of miles, lay back their heads and announce their arrival by joyously rattling their beaks; when the swallows have made their entry and hurry through our streets and pass our windows in sailing flight; when the lark appears as a dot in the ether and manifests its joy of existence by its song; then a certain desire takes possession of man. He longs to soar upward and to glide, free as the bird, over smiling fields, leafy woods and mirrored lakes, and so enjoy the varying landscape as fully as only a bird can do."

"Bird Flight As The Basis of Aviation"
—OTTO LILIENTHAL, 1891

Lilienthal's words eloquently express *the dream of man* through the ages. Even though it is now almost a hundred years since he wrote those words, they aren't any less true today—the dream has not faded. While men have been flying since then, mankind just hasn't been able to afford it. With the emergence of the modern ultralight however, the situation is changing. Low cost flight is now here for anyone to "enjoy the varying landscape as fully as only a bird can do." The following prologue is an example of the modern day realization of Lilienthal's quest.

The early morning air hangs still as the sun rises over the Allegheny Mountains. Fog is nestled snugly in the valleys, protected by the shadows of the ridges. The sky is crisp and clear. There's not a cloud in sight. It looks as though another super fine flying day has dawned on the Susquehanna Valley near Hummelstown, Pennsylvania.

It's eight o'clock on a Saturday; most people are still in bed. The top leaves of the maple trees show no signs of quivering. Cliff Goodall at the National Weather Service is predicting surface winds of five mph. Flight Service at Capital City Airport said that winds at 3000 were north-west at seven. This is definitely a day for flying ultralights — a day to fly for flying's sake!

No one is around to watch, as I emerge from my house in the country. I open the garage door with the push of a button — it's radio controlled, a marvel of the computer age. The car is backed out of the garage and parked in the driveway. A small "T"-trailer is hitched-up. The safety chain is connected — clink, clank, clunk — it's secure. An ultralight air vehicle is mounted on top of the trailer, with its wings folded neatly alongside the fuselage — if you want to call it that. This is where I keep the ultralight whenever I'm not flying it. Here, it's safe from "hanger rash", not to mention hanger rent. My 35mm camera is loaded into the trunk — you never know when a good opportunity might arise — along with a ten-gallon drum and two-cycle oil. Roberta, my wife, brings along my helmet and goggles, and a small lunch she prepared while I was fussing with the trailer. We're ready to go.

We drive to Skip's Mobil in Hummelstown to fill the drum with regular. It gives me a sense of security knowing I can get gas anywhere. We head out a couple of miles to the Keystone Radio Control Society flying field. I have special permission to fly from here, provided no R/C models are flying. The dew still clings to the freshly mowed runway. The windsock on the maintenance shack is limp. A solitary hawk glides by. The crows exchange caws on their "dawn patrol".

The ultralight is disconnected from its fastenings on the trailer. Metal clinks as we roll the craft to the ground. The wings are unfolded, the struts are connected, and various pins put in place. In no time at all, a flying machine is born — one that vaguely resembles a "real" airplane, but is obviously much simpler. Dacon surfaces are taut, sounding like a disco drum when tapped with a finger. The nose points skyward, as if to sniff the air. An ultralight airplane is poised for flight.

I preflight the flying machine with unhurried deliberation; examining every nut and bolt, tube and cable, fabric and hinge. I run my fingers along the edges of the prop, feeling for nicks, looking for cracks. The reduction drive belt is inspected for tension and wear. The engine mount is checked for integrity. Standing at the nose while facing forward, I move the stick. Left stick is left aileron up, and vice versa. Back stick is up elevator, and vice versa. I put my hands on the rudder pedals. Right is right and left is left. The throttle control cable looks good. The fuel line is clear and clamped. The tank is clean. I mix the gasoline and two-cycle oil like some introverted chemist. It must be in the proper ratio if the engine is not to seize or the plug foul. I pour the precious mixture into the tank. I look at the strut attachments, the landing gear and control system pulleys. I step back from the machine to assess its overall appearance. It looks good — it's ready to fly. The wheels are chocked.

I put on an extra sweater, then my jacket. It's a WWII fighter pilot style that makes me feel like a flyer when I wear it. A white scarf is tucked neatly inside — the prop would win if it ever streamed back — and the jacket zipped up. Next, I loop the loaded camera's strap around my neck.

Before I get into the seat, I stare at the surfaces of the wing for a moment. What a gift, this thing we call a wing. It's a marvelous invention. Thank you Otto Lilienthal. Thank you Orville and Wilbur Wright. Surely its as significant as the wheel, maybe more so. A wheel only rolls, but *a wing can fly*. It's a device that can bring you untold joy and pleasure, as it lifts you from the clutches of gravity for what seem to be all too brief instants of your life. If I had a choice, I suppose I'd rather be an eagle, where flying is a way of life. But as it is, I am a mere mortal man. But yet, I can raise myself above it all in my ultralight. I can see and experience what, until recently, was affordable by only a very few. It gives me goose bumps knowing that aerial delights can now be had by almost everyone on the earth. No longer is the sport of flying for the rich, thanks to the marvel of this ultralight wing.

Daydream aside, I crawl onto the seat and fasten the safety belts. Roberta hands me my helmet and goggles. I slip them on. The primer bulb is pumped a couple of times and I yell "clear". I check to see the ignition is off, and pull the engine through with the choke on. The cylinder walls should be wet now, as the liquid fuel is being vaporized. I crack the throttle, throw on the ignition and pull the starter. Suddenly, the engine comes to life. I let it idle awhile to warm up. The CHT is looking good. I rev it up and back off several times — the response is smooth, with no backfiring.

At idle now, I ask Roberta to remove the chocks. I am ready to go. Taxiing out to the runway, I recall my earlier hang gliding days and how hard we had to work, climbing up the hill for a short ride to the bottom. Those were fun times that gave me a kinship with the Wright Brothers — they started out gliding, you know. At any rate, I'm now turned into the normal takeoff direction — west. As I open the throttle, the staccato sounds of the engine harmonize with the song of the propeller. As the revolutions rise so does the tone, and the propeller bites into the cool air, turning faster and faster as I accelerate. The tail comes up, and in 75 feet I'm off the ground. It's a fantastic feeling to fly. The wind is in my face and the scenery is shrinking beneath me. It is good to be in the air again.

It is all I can do to contain my emotions, experiencing what is only a dream for most people — flying. Every time I take off, I am like a child with a new toy at Christmas. Every flight is different — filled with new delights. I can hardly believe that something as thin as the life-sustaining air I breathe can exert enough force to lift this machine and my ungainly body off the ground. At last, I have defied the law of gravity and won. If only Sir Issac Newton could see me now. I am ecstatic over the sensation. I scream and yell at the top of my lungs to let those ground-bound souls below know how wonderful it is — *to fly*. Get out of bed you sleepy heads.

I want to share this dream come true with everyone I see. If you like magnificent views, try one from the sky. The earth is truly beautiful from this lofty perch. The randomness of Mother Nature shows a kind of order from the air. I'm overcome with a sense of ultimate power. I feel euphoric and somehow god-like. It must be the diminished size of objects below. They all fall into insignificance. There is nothing I cannot do. *I can fly*.

After a straight climb to 300 feet, I crank in a 180 and head back over the field. Full power elevates me at about 750 fpm. I'm already at 1000 feet now. Another 180 puts me back into the wind for a little steeper climb. As I pass over the field, I rock my wings, a signal to Roberta that I'm on my way — heading for the Tuscarora Mountain and some soaring. There should be some hang gliders there today.

I throttle back, setting up a lazy cruise of 40 mph — I'm not in a hurry. The engine is barely audible. Now I can relax as I set a course to intersect the Susquehanna River. There's George's house off my right wing tip. Looks like he's washing his car. I circle him, and he looks up waving. I know he'd like to fly; most people I talk to say they dream about it. But I'm doing it.

There's the river. A gentle right turn and I'll be right over it, heading north. What a beautiful view. The pastures along the river seem to be smiling up at me. The water glistens as the rays of sun clear the ridge tops. The low-lying fog is melting away. I breathe deeply. It's great to be alive — *to fly*.

There's the Dauphin Narrows — it's easy to see why they're called that. Up ahead I can see the Clark's Ferry Bridge, where the Juniata joins the Susquehanna. I must now follow the Juniata, northwest to Thompsontown Station and the "Tusc". I think I'll take a few photos.

A half hour goes by. Gosh, I forgot my lunch. No matter, I'll land at Mifflintown and grab a bite. I'm getting closer to the mountain, I'm almost 800 feet above its 1200 foot peak. Approaching carefully on the windward side, I throttle back to idle and descend. At about 300 feet above the ridge, my body senses the slope lift as a gentle push from the seat. My descent is checked and I'm soaring. I can shut down the engine now; it can take a breather.

All I can hear is the sound of the wind in the wires as my wings slice through the late morning air. I see a few hang gliders setting up the "chute" at the top of the mountain. Not everyone is into powered ultralights. Hang gliding can be a lot of fun.

I hear the shrill cry of a hawk — he's soaring beside me — off my left wingtip. He comes in a little closer, as if to check me out. He accepts me. We soar together for 15 minutes. I'm at peace with myself and Mother Nature. This is ecstacy — to fly.

Prologue

There goes a hang glider, then another and yet another. Conditions are marginal, but the new gliders should be able to soar. They sure have come a long way since the old standard Rogallos of the early '70s.

Suddenly, my jubilation turns to terror — I'm sinking! I've lost it. I'm out of the ridge lift and heading straight for the trees. Is Mr. Newton getting his revenge?

No matter — I simply start my trusty engine and climb again. How sweet it is. That little "two-banger" has "saved" me countless times before. If I had been in a hang glider, I would have eaten a pine tree.

I've caught the lift again and can shut down the engine. It's silent once more. I feel like the center of the universe — it revolves around me at my command. I am awed by the specter of my flight as I bank into a vertical turn. My illusion vanishes in a rush of wind and centrifugal force. The mountain tilts; the river looks like a whirlpool around the inside wing.

I have known the beauty of flight from my experiences in Piper Cubs and Cessna 150s, but this is unlike anything I've ever felt in an airplane. I'm not encapsulated in an aluminum and plexiglas shell. There are practically no gauges. My engine burns only a gallon, or so, of regular auto gasoline an hour. This is powered flight in its simplest, purest form. I can feel the wind against my body. I can feel the forces of flight. I can hardly believe this is real. *At last I can really fly . . .*

This book is an attempt to explain the basics of flying the new ultralight air vehicles. It is an attempt to give you some insights into what it's like and how it's being done. If you are a licensed pilot, it will provide you with a good review of the basic piloting skills you may have forgotten; of navigational procedures you may have discarded for VORs, transponders and DMEs. In addition to operation from land, the book also describes how ultralights are operated from water and snow, unlocking the door to vast recreational areas. The special requirements of operating in the airspace are also included, along with the latest FAA rules and advisory circulars. The book should fulfill a long standing need for those who wish to fly ultralight air recreational vehicles — pilots and potential pilots alike.

"He longs to soar upward and to glide, free as the bird, over smiling fields, leafy woods and mirrored lakes, and so enjoy the varying landscape as fully as only a bird can do."

1
The Principles of Ultralight Flight

Unlike surface conveyances which are restricted to two-dimensional movements, aircraft are free to move about in all three dimensions. In addition to moving straight ahead, side-to-side and up and down, an ultralight is capable of making three fundamental rotations, one about each of its three axes. The three axes/rotations are pitch, roll and yaw.

The Three Reference Axes

The pitch, or lateral, axis is an imaginary line running through the center of gravity, parallel to the span of the wing. Rotation about this axis produces changes in pitch attitude, i.e., nose up - nose down. Pitch attitude, and therefore angle of attack, is controlled by the elevator, taking on a fixed degree, depending on elevator setting.

The roll, or longitudinal, axis is an imaginary line running from the nose of the aircraft through its CG, parallel to the direction of flight. Rotations about this axis are governed primarily by the ailerons which, when deflected,

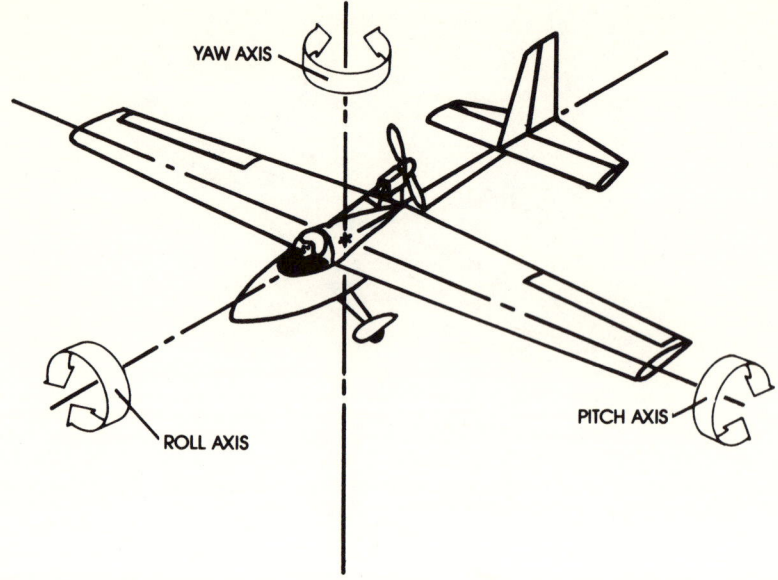

Fig. 1-1. The reference axes of an ultralight.

establish a *rate of roll* rather than a fixed bank angle. Once the desired bank angle is achieved by aileron deflection, the ailerons are neutralized or even reversed, to maintain the desired bank angle. Constant aileron deflection would cause the aircraft to continue rolling.

The yaw, or vertical, axis is an imaginary line passing through the CG perpendicular to the other two axes. Rudder controls yaw, a given rudder deflection resulting in a particular *yaw rate*. Once a yaw change is accomplished, the rudder should be neutralized. Holding rudder will cause a continuous yaw.

THE BASIC ELEMENTS OF FLIGHT

An aircraft is considered to be under control when the pilot is able to direct its course at will. This is the case as long as the aircraft remains unstalled, and is moving in the desired direction at the selected altitude. The pilot is kept informed of these conditions by the three primary flight instruments: airspeed indicator, altimeter and compass. These are basic to all flying, and tell the pilot what *state-of-flight* his aircraft is in.

As necessary as instruments are however, no pilot should rely solely on the information they present. After all, they are mechanical and they could fail. More important than that though, every pilot must develop a sense for *"the-feel-of-flight."* He should be able to judge his flight condition by the audio and visual messages received by his brain and body — the instruments being used as cross references and "fine tuning" devices.

The Principles of Ultralight Flight

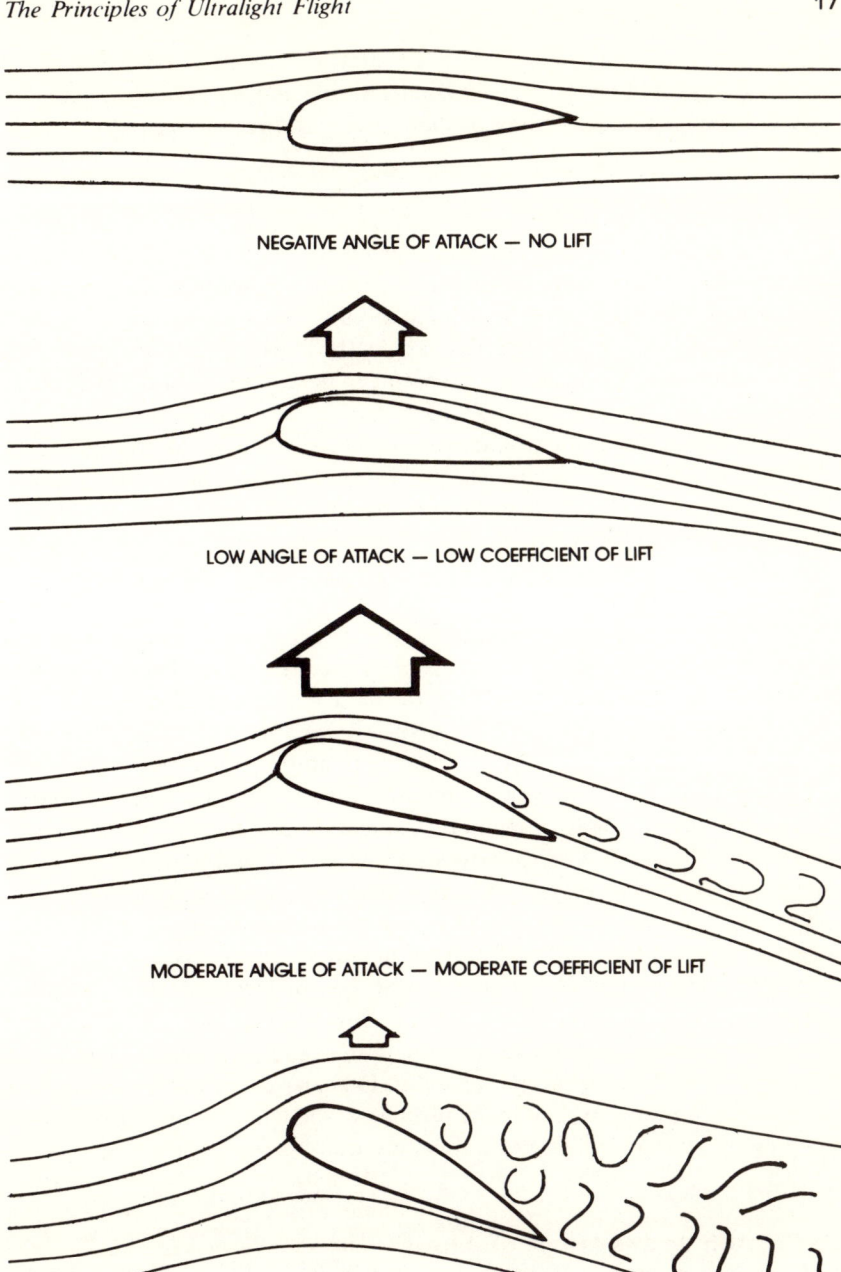

Fig. 1-2. How the airflow changes over a wing at varying angles of attack. The coefficient of lift increases with angle of attack until the airflow separates from the upper surface, causing the wing to stall.

Recently, an inexpensive angle of attack indicator has been made available. It should give an absolute indication of the *degreeness of flight* and lead to a better understanding of flight and stall prevention.

The First Element of Flight

First and foremost, *the essential element of flight is that the wing be installed.* This means the angle of attack must be below the stall angle. It also implies that the airspeed must be high enough to avoid stalling at a given angle of bank. (The higher the bank angle, the higher the stalling speed, but more on that later.) The proper airspeed can be related to sound and feel — *the sound and feel of flight.* Under normal cruising conditions, your ultralight will produce a particular sound — the sound of the wind flowing past the wires and struts and over the wings. Under normal cruise you will also be able to feel a certain stick pressure with your hand. Another clue to your *degreeness-of-flight* could be the flagging motion of your pant legs (in open air types) — the faster they flap the higher your airspeed and the lower the angle of attack.

The stick can be a good indicator of the *degreeness-of-flight* because it is where the feel of the controls is transmitted to the pilot's hand. Since the control surfaces are in the airstream, the higher the airspeed, the harder the control forces become. Under high airspeeds, the stick feels "hard," but only small movements are required to make the aircraft respond. As the airspeed diminishes and the wing approaches stall, the stick gets "soft" and large movements are required to maintain control.

One must be careful interpreting stick feel. On most airplanes the tail is in the propwash. This means the tail could experience a greater airflow, increasing tail control surface effectiveness at near-stall airspeeds. The ailerons, of course, would "see" only the true airspeed and become ineffective or useless. The propwash induced harder feel of fore and aft stick (pitch control) could be interpreted as an indication of a high *degreeness-of-flight* which, in reality, wouldn't exist near the stall example cited here. The astute pilot will be aware of this false indicator. On canards, of course, the rudder and elevator surfaces are not in the propwash and will see only true airspeed, just like the ailerons, giving no false stick feel.

Pitch Attitude and Angle of Attack

Another very important visual cue in flying is that of aircraft attitude — its orientation with respect to the horizon. Cruising flight entails a particular aircraft attitude with respect to the horizon. Depending on your perspective, you can determine whether the nose is up or down, and if the wings are level. For straight and level, unaccelerated flight this is fine, but attitude alone cannot be used to determine your ultimate *degreeness of flight*. Your main concern is that the wing be unstalled.

Gravity and Acceleration

You can feel the effects of gravity and acceleration directly through your posterior anatomy. Under normal straight and level unaccelerated flight, you experience a one-'g' load (your weight), just as you would sitting in a chair in your living room. Put an airplane in a 45° banked turn however, and your weight will increase 41%, which you'll feel through *the-seat-of-your-pants*. If you experience a negative 'g' situation, the seatbelt will tug at your gut. If your turn is less than coordinated, you'll be able to feel a side force, much like that felt in a car going around an unbanked turn, tending to slide you out of your seat. When a pilot is aware of and can sense accelerations with his body, he can make appropriate control corrections to restore coordinated flight.

The Second Element of Flight

Secondly, it is essential that you have *altitude sufficient for safe flight*. An altimeter is essential for this information, and the higher you go, the harder it is to judge. Once a power setting is established and the airplane trimmed for straight and level flight, altitude should remain constant, except for updrafts and downdrafts. Altitude is essential for safe flight. You should have enough to glide to a field in case the engine quits, and certainly enough to recover from a stall. Flying at tree top level might be exciting, but it is illegal over congested areas, and can be quite dangerous. Altitude gives you a certain margin of safety.

THE THREE DEGREES OF STABILITY

In order for an ultralight to be "pleasant" to fly, it must possess certain handling qualities. It should be capable of being flown "hands-off" for a short period of time, which implies that it is stable.

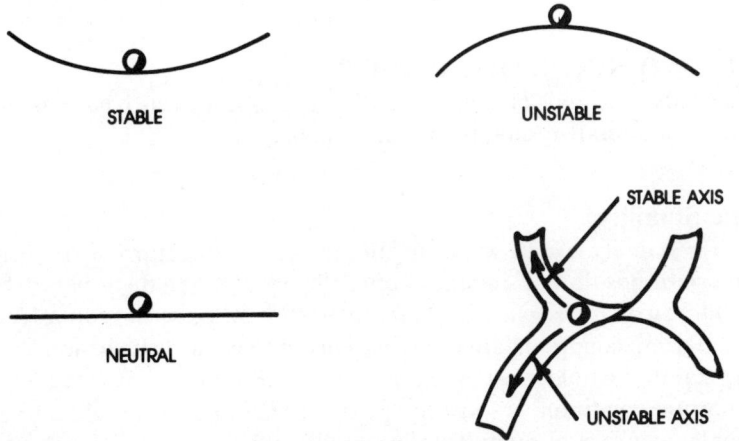

Fig. 1-3. The three degrees of stability. It is also possible for a system to be stable on one axis and not on another.

Positive Stability

Positive stability can be defined broadly as the ability of an aircraft to return to its original equilibrium flight condition, after having been disturbed from that condition. Equilibrium (balance) exists when all the forces acting on an aircraft, or any body, are balanced by equal and opposite forces. In order for an aircraft to be in equilibrium, lift must equal weight and thrust must equal drag. If these forces are not balanced, the aircraft will accelerate either by changing speed or changing direction.

A marble in a bowl would be a good example of positive stability. If it were disturbed from its stable equilibrium position at the bottom of the bowl, it would tend to return there once the disturbance is passed.

Neutral Stability

Neutral stability is displayed whenever a body in equilibrium is disturbed, and shows no tendency to either return to its original position or condition, nor move further from it. It has no preference for its state of equilibrium — it is neutrally stable. A marble on a flat horizontal surface would be neutrally stable. It stays where it is put, and has no tendency to return to its original position, or deviate from the new one.

Negative Stability

A body exhibits negative stability when, after having been disturbed from equilibrium, it has a tendency to move further away from its original condition or position. Negative stability has no place in an airplane! A marble sitting on top of a bowling ball would be negatively stable. Once disturbed from its equilibrium condition, it accelerates away from it, never to return.

THE TWO KINDS OF STABILITY

There are two kinds of stability: static and dynamic, which have to do with the way an aircraft responds to a disturbance.

Static Stability

Static stability exists when an aircraft tends to return to its original equilibrium position or condition after the passage of a disturbance. Static instability would be where an aircraft would continue deviating from its original trim condition after having encountered a disturbance. Neutral static stability would exist when an aircraft exhibits no preference for an equilibrium condition. It's just happy flying sideways as straight ahead, for example — a case of neutral static stability in yaw. Most ultralights today appear to exhibit static stability about all three axes — some more than others.

The Principles of Ultralight Flight

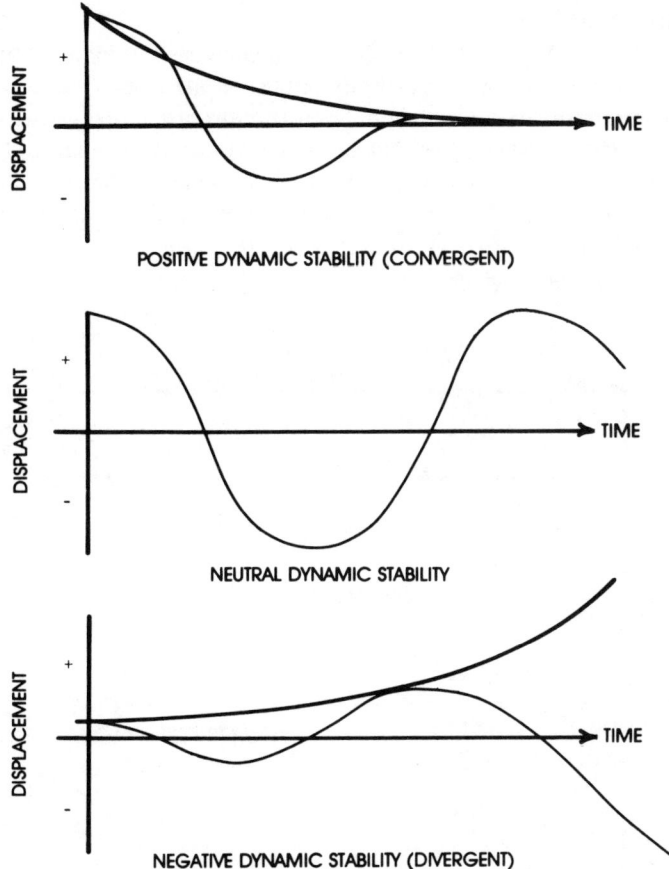

Fig. 1-4. The three degrees of dynamic stability.

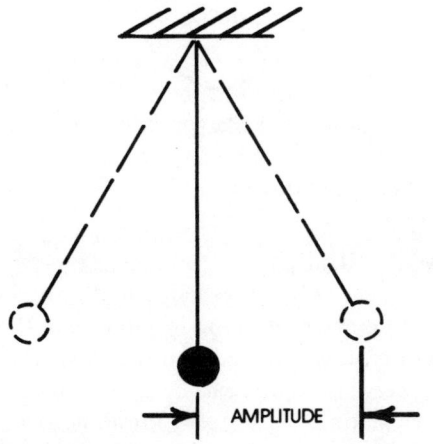

Fig. 1-5. A free pendulum is both statically and dynamically stable. It's amplitude (displacement) decreases with time, after it is let go.

Dynamic Stability

Dynamic stability is a little more complicated case. It involves oscillation — back and forth rotations about either or all axes — that gradually decrease or damp out with time, until the aircraft returns to its original attitude or trim condition. An unrestrained pendulum is dynamically stable, since the amplitude of its swing lessens with time. If the pendulum were immersed in water, the time period for it to stop oscillating would be lessened. In other words, the water damps the motion. A large vertical tail would tend to increase your damping.

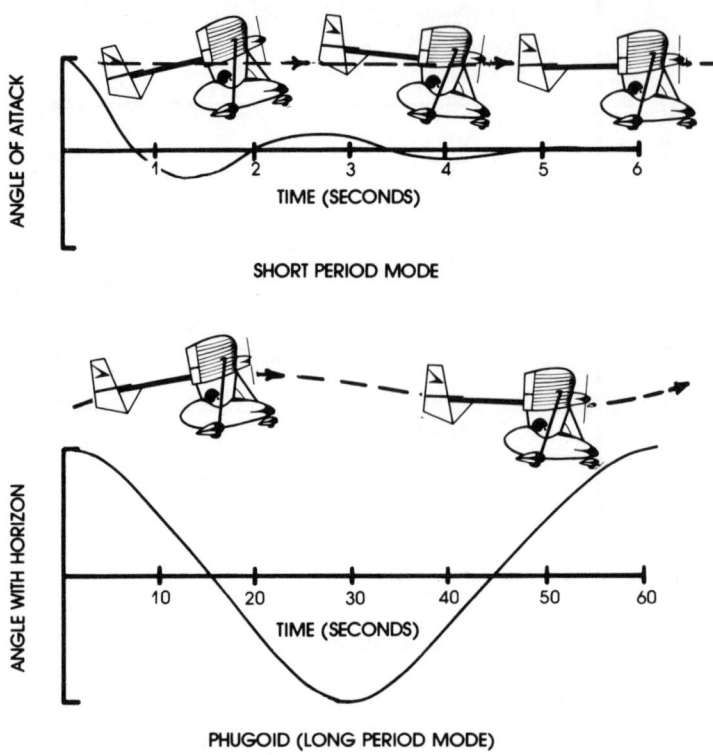

Fig. 1-6. The two longitudinal dynamic modes.

Overall Stability of Ultralights

Most ultralights appear to be stable in pitch, the ones with horizontal tails or canards exhibiting greater damping in pitch than the flying wing types. Basically, once an aircraft is trimmed, it tends to seek and maintain a fixed airspeed. It will either climb, descend or maintain altitude depending, of course, on throttle setting. A cruising aircraft disturbed in pitch will eventually return to level flight after going through a gradually decreasing "roller-coaster" type ride, provided it is dynamically stable in pitch.

Most aircraft are marginally stable in the lateral and directional senses. An aircraft cruising along in level flight can return to level flight all by itself, after an encounter with a side gust, so long as practically no turn is established. Once a turn sets in though, the original state of equilibrium is upset, due to the generation of centripetal force. The situation worsens as the wing on the outside of the turn travels faster, generates more lift and increases the bank angle still more. If left unchecked, the nose will drop and a spiral dive will develop. Fortunately, this tendency is not noticeable in normal pilot-controlled flying. Also, the proper combination of dihedral and vertical fin can make an airplane spirally stable (as is a free flight model airplane). Trouble is, this extra stability also lessens the aircraft's roll control.

Trim

An airplane is normally designed so that no stick or rudder pedal control forces need to be exerted during cruising flight in a calm, for a particular CG location. In other words, the flying surfaces are rigged so the airplane will fly straight and level with your hands off the stick and feet off the rudder pedals. If your airplane has the pilot situated at other than the CG, or you wish to fly faster or slower than cruise, then the trim will need to be changed.

Whereas most light planes have a cockpit adjustable pitch trim system, ultralights do not. Since they are single seaters flown typically only by the owner, there seems to be little need. However, some trim adjustments may be required for stick force free flight. A ground adjustable stabilizer or small metal tab at the end of a control surface, or even a spring-loaded stick, are the methods of trimming an ultralight.

The only way to determine if you need to re-trim your ultralight is to fly it. Get to a safe altitude and set up a cruise throttle setting and see if the ship holds a constant altitude, hands off. If it wants to descend, the nose is heavy. The stabilizer leading edge needs to be brought down a tad, the stick needs spring loaded to the rear, or the elevator tab bent down. After a *small* adjustment is made, take it up again and check the trim condition once more. Consult your flight instructor or dealer first, before making adjustments. If your airplane is out of trim, the CG may be incorrect, possibly resulting in longitudinal (pitch) instability.

Aileron and rudder trims are less likely to need adjustments. However, if a side force on the stick is necessary to keep a wing up or you need to hold rudder pressure to head in the direction you want to go in a calm, the aircraft may be inproperly rigged, and out of trim. Consult your dealer or flight instructor for advice.

Climbing and Gliding

Climbs and glides are basically straight and level flight with power either increased or decreased from that required to maintain altitude. What we are talking about here is basic to understanding aircraft control — *the throttle is really what makes you go up or come down. It is not your speed control.*

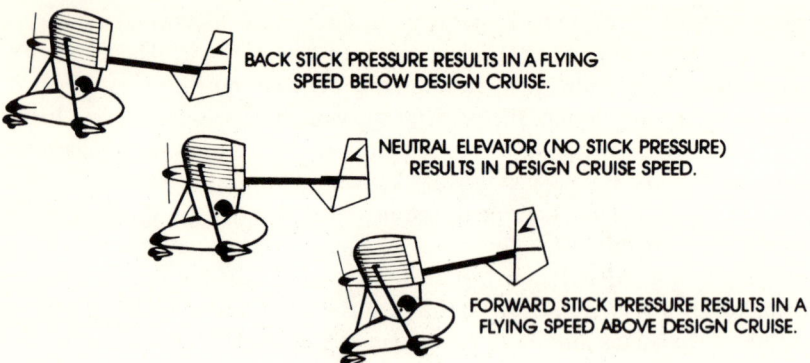

Fig. 1-7. Your trim determines your flying speed.

Under normal conditions when the throttle is fixed, such as during a climb, cruise or approach, *the airspeed is controlled by the stick!* Sure, pulling back on the stick while cruising will make you go up, but only momentarily. The nose high attitude (higher angle of attack) will cause the airplane's drag to increase, reducing the airspeed (you may need to advance the throttle and/or lower the nose to prevent a stall).

Now you say, "but the throttle controls the airspeed whenever the altitude is fixed." Guess again. Granted, if you advance the throttle your speed will increase — but only if you also retrim the airplane by holding forward stick! Otherwise, *the airplane will maintain its original airspeed* while commencing to climb.

While gliding throttled back, as in an approach to landing, it is important to remember that *pulling back on the stick will not stretch your glide* — it will actually steepen the glide path and maybe even stall the aircraft. This seems contrary to what you might at first believe, but a little thought will reveal the reason: the higher angle of attack caused by the back stick causes an increase in drag, degrading the maximum glide ratio you were at or near during the approach. The angle of attack simply must be maintained below stall at all cost, for without it, you'll fall out of the air. When near the ground, keep the stick forward and maintain your airspeed. It is far better to reach the ground with a little too much airspeed, than to stall out at 25 feet!

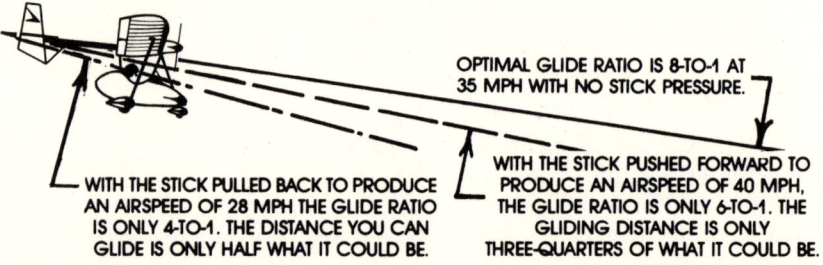

Fig. 1-8. Stick pressure during a glide (power-off or idle) will control your airspeed and glide ratio.

2
Preflight Inspection

PREFLIGHT — A thorough walk-around inspection of the air vehicle, to be performed before each flight. It should be done systematically and habitually, and always in the same sequence.

First of all, assemble the vehicle for flight, being certain the ignition is off. Begin your inspection at the nose and walk around counter-clockwise to the right. In general, you should look at the following items.

NUTS/BOLTS — Located, not stripped, secure, safetied and not compressing tubing.

POP RIVETS — Tight. No sheared heads.

CABLES — In tension, no frays or broken strands. Thimbles and Nicopress swages secure. Not wrapped around tang or bolt.

FABRIC — Secure around perimeter of flying surfaces. No tears or wrinkles. Tight and velcro sealed.

TUBES/STRUTS — Straight (or in manufactured configuration), no bends, kinks or cracks. Pinned or bolted, and safetied.

CONTROLS — Stick, pedals and levers operating freely, secure and safetied. Control surface hinge pins in place and safetied. Be sure controls respond properly to stick movements; moving stick toward an aileron causes it to go up, and moving the stick back causes the elevator to go up for aft tails and down for canards.

LANDING GEAR — Tires inflated and in good condition, wheels rotating freely and safetied. Control cables or struts attached to nosewheel or tailwheel.

PROPELLER — Clean surface with no cracks or missing pieces. Remove insects, mud and stones. Bolted to hub and safetied. Balanced.

ENGINE — Bolted and safetied in its mounting. Check mounting for cracks. Carb and air filter clean and secure. Spark plug cable, throttle linkage and fuel line secure and safetied. Needle valves tight. Muffler sound and secure.

FUEL LINE — Clamped and clean. Drain if dirty inside. Clear air bubbles before take-off. Filter clean and properly secured.

FUEL TANK — Full. No water or other contamination. Proper fuel oil mixture. Cap tight.

INSTRUMENTS — Positioned, secured and connected. Airspeed sensor properly aligned for correct readout.

IGNITION — Battery charged. Electrical connections secure.

SEATING — Harness attachment secure. Webbing not worn. Seat secure. Seatbelt anchored and not worn.

Back away from the airplane and sight for overall appearance and trueness. Be certain it is trimmed for your weight, including clothing and accessories. Do not deviate from manufacturers recommended assembly and rigging procedure. Any alterations could seriously affect stability and flying characteristics.

NOTE: As time goes on and the airplane is used, moving parts wear and materials breakdown from exposure to the elements, as well as from flight stresses — watch for gradual deterioration. Preventive maintenance is a must for aircraft. If your airplane is newly assembled or repaired, check it thoroughly before attempting flight. HARD LANDINGS DESERVE A COMPLETE AND THOROUGH INSPECTION!

A complete preflight inspection is an absolute must before each flight, and it is not to be taken lightly. If something on the ultralight looks questionable, DO NOT FLY! Those who overlook the details usually won't live to tell about their haste.

Since all ultralights are not the same, it is virtually impossible to publish an exact preflight procedure for each model. Each manufacturer is responsible for this. The author however, presents the sequence he uses on his ultralight, for your information.

Preflight Inspection Procedure Checklist
1) Cockpit Area
    1. Control stick lock — Remove.
    2. Ignition switch — Off
    3. Controls — Aileron moves up on wing panel toward which stick is moved. Stick moved back toward the tail produces up elevator and

5) Tail Group
   1. Stabilizer — Hinges secure. Elevator moves freely.
   2. Vertical Tail — Hinges secure. Rudder moves freely.
   3. Bracing Wires — Properly tensioned and securely anchored.
   4. Tail Wheel — Tire good. Mechanism sound.
6) Right wing trailing edge
   1. Aileron — Hinges secure. Aileron moves freely.
7) Right wing strut
   1. Strut to wing anchor point — Secure and safetied.
   2. Strut to fuselage anchor point — Secure and safetied.
   3. Strut — Free of kinks, bends, dents, and scratches.
8) Landing gear
   1. Right Wheel — Tire properly inflated. Wheel secure and safetied.
   2. Left Wheel — Tire properly inflated. Wheel secure and safetied.

Starting and Taxiing   29

3
Starting and Taxiing

Before the beginning of each flight, an aircraft's engine must be properly started and operated as the aircraft is taxied (maneuvered on the ground). It is essential to become proficient in taxiing and throttle use before flight can even be considered. Consult your flight instructor before going any further.

The Landing Gear
There are two primary types of landing gear: taildragger and tricycle. The taildragger has two main wheels in front and a smaller wheel near the tail. The tricycle arrangement has two main wheels behind the CG, and a nosewheel in front. A steerable nosewheel or tailwheel turns with the rudder. The tricycle landing gear offers superior ground handling characteristics and could allow the wing to be at a negative lift angle while on the ground, a tremendous asset when operating in winds. The taildragger configuration is not quite as good. It is possible to "ground loop" (the tail swings around to position itself in front of the nose) and is more affected by gusts picking up a wing, because it is at a lifting angle of attack.

The Throttle
The throttle controls the speed of the engine and therefore the thrust of the propeller. Pushing it forward opens the carburetor, allowing more fuel and

Fig. 3-1. A two seater ultralight trainer with dual controls and tricycle landing gear.

air to be drawn into the cylinder(s), increasing the power. Pulling it back (closing the throttle), does just the opposite. Proper use of the throttle is essential to good taxiing technique.

Starting the Engine

Before starting any engine, be sure the wheels are chocked, or the aircraft is otherwise restrained from movement. *The propeller area is to be clear of spectators as well.* Also see that the fuel tank and gasoline lines are full. If you have a transparent tank, check for water and other contaminents in the bottom — water is heavier than gasoline and it stops engines when it gets to the combustion chamber! If you have an opaque tank, it should be drained and cleaned to minimize fuel line contamination.

After all of the above is completed, see that the ignition is off. Crack the throttle, i.e., open it to the idle position, so the engine doesn't race when started. Prime the engine with a squeeze or two of the primer pump, and turn it over a few times by hand to draw some fuel into the cylinder(s).

Now, look around the area a full 360 degrees to see that no one is near the propeller. Turn on the ignition and yell "clear" or "clear prop" and pull the starter — the engine should start. If it doesn't, try a little more prime. If it still doesn't start, or even pop, perhaps the ignition system is at fault. Check the wiring and inspect the spark plug for fouling. The engine may be flooded, or have too much fuel in the cylinder(s).

If the engine is flooded, a soaked spark plug as well as a heavy gasoline odor will tell you so. It is necessary to turn off the ignition and open the throttle wide. Now, turn the prop backwards a few revolutions to draw out

Starting and Taxiing 31

Fig. 3-2. A throttle lever mounted for left hand control.

the excess fuel, and clean off the plug. Replace the plug and try the starting procedure once again. If you still encounter starting problems, consult a copy of Ultralight Propulsion, published by Ultralight Publications, Inc.

Hand Propping an Engine

If your engine has neither a recoil or electric starter, you will be forced to start it by hand. In the early days of flying, this was the only way to start an airplane engine. Anyway, it can be quite dangerous and requires an assistant. For the correct procedure you should consult Ultralight Propulsion.

Keep Your Hands on the Throttle and Switch

A habit you must get into is that of keeping your hand on the throttle, not only on the ground, but in the air as well — you could need it at any time. Also remember to keep the ignition switch off at all times except when starting and running. If the switch were left on while turning the prop by hand during the priming period, the engine could "catch" violently striking the hand or arm, perhaps breaking it.

Keeping your hand on the throttle while starting will also tell you that it's not wide open, or closed. If opened wide and the engine starts, the airplane would lunge forward and could cause all sorts of problems. If closed, the engine would start alright only to quit after the prime was burned.

Warm the Engine

After the engine is running, set it at idle rpm and let it warm up to the manufacturer's suggested temperature. Never attempt flight with a cold

Fig. 3-3. A side mounted control stick on a single seater ultralight. It controls both pitch and roll.

engine — it is more likely to quit than a warm one. After the engine is up to temperature, point the aircraft into the wind and either chock it or have someone restrain it from moving. Now, take the throttle up and down a few times to see that it advances smoothly and doesn't backfire on the way down. If you have dual ignition, check each circuit by shorting — a small rpm drop should occur on each. If not, one or the other or both is defective and will have to be checked before flying.

Taxiing — Maneuvering an Aircraft on the Ground

The ultralight student is often surprised by the difficulty he has in learning to "drive" his aircraft on the ground. It doesn't handle quite like a car! The main reason for this is that it's a flying machine first and a ground vehicle last. The controls are designed to function best at speeds above the stall, while most taxiing is done below the stall. At any rate, after some practice, taxiing becomes second nature, as you learn to feel the aircraft's responses to various control inputs at various speeds.

Taxiing techniques vary from ultralight to ultralight, but some basic rules apply. In flight, there is a strong airflow past the rudder, making it more effective. While taxiing at low speeds, the good airflow doesn't exist, and control response is slow, especially if there is no steerable nose or tailwheel. As airspeed increases, the rush of air past the rudder makes it more effective. If the rudder is in the propwash, turn response at low taxi speeds can be quickened by a momentary blast from the propeller.

On taildraggers, forward stick (i.e., down elevator) can be used to lift (assisted by momentary prop blast) the tailwheel off the ground, and allow

Starting and Taxiing

the rudder to swing the aircraft around. Once a turn is started at higher taxi speeds, it wants to continue and requires anticipation and opposite control before straightening is desired. Remember, taildraggers are prone to ground looping.

On tri-geared ultralights, backstick (up elevator) will lift the nose, allowing the rudder to swing the aircraft around, as the throttle is momentarily advanced. Tri-geared aircraft are easier to taxi than taildraggers and don't require such intense anticipation and vigilance.

Aileron can also be used while taxiing at higher speeds, but gives a response opposite to that in flight. To turn left, the stick is pushed right (lowering the left aileron and raising the right) allowing adverse aileron yaw to develop a left turning tendency.

High Speed Taxiing

Before you can even consider heading for the wild blue yonder, you must become intimate with the high speed taxiing characteristics of your aircraft. The method the author uses to familiarize himself with a new ultralight is presented. It is a slow, conservative approach, but necessary to really get the feel of a new plane.

First of all *clear the area* by checking to see that there is no other traffic either taking off or landing. It's a good idea to have an assistant watch out for traffic while you are practicing your taxi technique — just in case. If it's calm out (calm conditions are certainly recommended) you can head in any direction, otherwise, always aim directly into the wind.

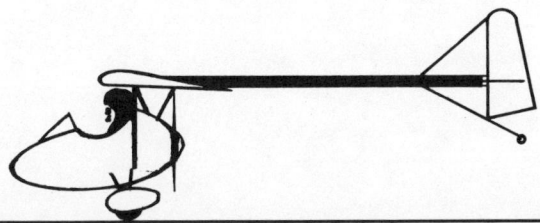

Fig. 3-4. A high speed taxi in a taildragger has the tail flying. This is the attitude assumed prior to liftoff and during a "wheel" landing.

Your first goal in learning to takeoff is simply to practice high speed taxiing. Pointed down the runway, advance the throttle gradually to the point where you're rolling along at around 5 mph. You may have to give a burst of full-throttle to get the initial motion started, but throttle back once you begin to roll. Immediately start feeling the craft out. How does it respond to fore and aft stick? Side-to-side stick? Rudder pedals? Be able to taxi at this speed in a straight line, right down the center of the runway. Make turns to the left and right — go 90, 180 and 360 degrees around.

Try the "turning-on-a-dime" technique. With a momentary blast from the propeller, raise the tail (on a taildragger) or raise the nose (on a tri-gear) while simultaneously kicking in either right or left rudder. You'll be amazed at how quick a turn can be made.

After you become proficient at 5 mph, advance the throttle to the point where you roll, say, 10 mph. Practice, as above, going straight down the runway and turning. Your turns will have to be wider and more gradual at this speed. The tail control surfaces should show a hint at being effective. Try taxiing with the tailwheel or nosewheel off the ground and on the ground. Taxi at this speed until you are comfortable with your reactions and those of the machine.

Now advance the throttle to roll 15 mph and repeat the above procedures. Here, the ailerons should start being somewhat effective but remember, they tend to work opposite. The adverse aileron yaw works at these below stall speeds, so left stick causes a right yaw and vice versa. Practice taxiing at 15 mph until you are comfortable with the aircraft.

20 mph is your next high speed taxi practice area. Here, you are approaching flying speed, and the airplane could become "light-on-its-feet." Things start happening fast. Only small rudder pedal inputs are needed to keep the aircraft rolling down the runway center. Sudden control applications are to be avoided. A yank back on the stick could cause you to leap into the air, stall, and come bouncing down. A quick application of rudder pedal could cause the outside wingtip to strike the ground, resulting

Fig. 3-5. Several ultralights waiting on a taxiway, while a Hawk comes in on final approach with flaps.

Starting and Taxiing 35

in damage to the wing and/or landing gear. Practice taxiing at 20 mph until you are proficient at it.

At taxi speeds above when the tail is effective, and below the stall speed, you'll find that back stick slows the airplane, while forward stick lets it accelerate. This is an important feature in using your wing as a brake to shorten your landing roll. Back stick also gives tailwheels more authority, while forward stick does the same for nosewheels — the extra weight (down lift) imposed on these wheels is the reason.

Now we are to the point of taxiing right below or at the stall speed. Here, the aircraft is almost ready to fly, and the controls will start responding that way. The wheels will be exerting very little weight on the ground and not offer much resistance to sideways motion. This is particularly true of taildraggers. On tri-geared aircraft however, holding forward stick could put a lot of negative lift on the wheels, depending on the particulars of the design, making it hug-the-ground.

At this speed, you can probably "pick-up-a-wing" with aileron, but watch out for the adverse yaw in the opposite direction and be ready for appropriate applications of rudder. Practice taxiing right below stall speed until you are thoroughly comfortable "driving" your airplane on the ground at high speed. The next step will be actually lifting off the ground, but you must prepare for it by perfecting your high speed taxi runs.

Crosswind Taxiing

Up until now, we've concerned ourselves with ground handling only in calm or straight into the wind situations. When you have to operate in a crosswind or turn away from the wind, you add certain complications to your taxiing.

Crosswind taxiing requires paying strict attention to wind directions with respect to the aircraft. A right crosswind could easily lift the right wing, causing the left wing to strike the runway, resulting in a ground loop to the left. Dihedral aggravates this effect because the underside of the wing "catches" the crosswind. The more dihedral your aircraft has, the more susceptible it is to crosswinds. If your airplane has minimal dihedral or none at all, it should handle crosswinds a lot better. The Wright Brothers used this knowledge to their advantage by actually rigging their "flyers" with *negative* dihedral. Their gliding experiments on the windswept dunes of Kitty Hawk were safer for this arrangement.

Provided you are not too far off the wind direction and your dihedral is low enough, your airplane will probably weathercock. This is a tendency to automatically nose into the wind, due to the side force on the tail created by the crosswind. As long as you can keep the windward wing from lifting up, you'll be able to handle some crosswind. This can be done by holding stick into the crosswind, while using rudder to maintain the desired ground track.

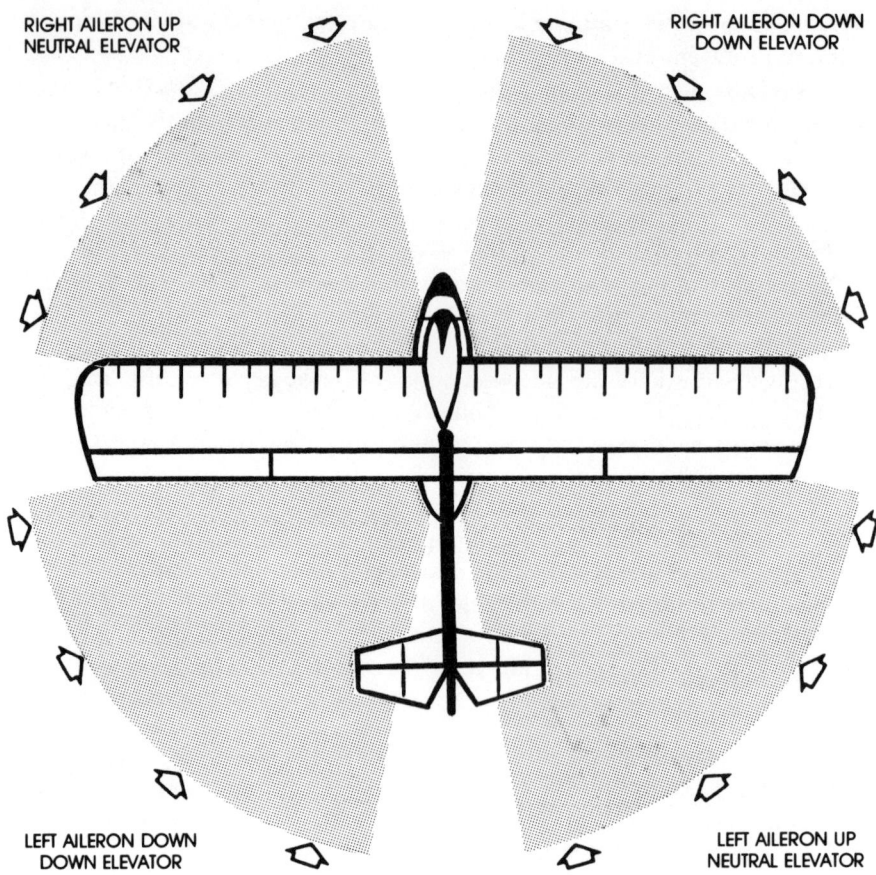

Fig. 3-6. Taxiing diagram indicating control positions required in various crosswind directions.

To minimize the possibility of being tipped over in a crosswind, the following stick positions are recommended for aileron equipped aircraft:

Right/Front Wind — Right Stick
Right/Rear Wind — Left Stick
Left/Front Wind — Left Stick
Left/Rear Wind — Right Stick
Nose/Tail Wind — Neutral Stick

If you analyze these positions and the winds, you'll see the intent is to deflect the ailerons so that wind pressure helps keep the windward wing down.

Practice taxiing until you are extremely confident and comfortable with the airplane's handling. Remember, when you land, you'll be in a high speed taxi situation and you'll have to be able to make a successful transition from flight control response to taxi responses. You must be ready for it before you do it!

Establishing Wind Direction

The common device used to tell what direction the wind is coming from is called a wind sock. It's actually a cone of fabric attached to a pole on top of a hangar or planted in the ground, where the wind can reach it unobstructed.

At first, the proper takeoff direction as indicated can be confusing, but if you remember to takeoff into the loose end of the sock you'll be right every time.

"The throttle is really what makes you go up or come down. It is not your speed control."

4
Crow Hops

After you become proficient at your high speed taxiing technique, you should be ready for the next phase of your flight training. It is now time for you to "unstick" from Mother Earth. NOTE: While the crow hop technique is used by the author to check himself out in single seaters, it is recommended you take dual flight instruction in a two-place ultralight. It's the best way to learn how to fly.

One of the most sensible methods the author has come across in checking out in a new ultralight, is known as the "crow hop." And, it should prove equally valuable, if not more so, in learning to fly. It consists primarily of skimming the ground and flying over it at heights from a foot to say, within a wingspan or so of the ground.

A crow hop is simply an extension of the high speed taxiing you accomplished earlier. You won't need full power flying this close to the ground, since you are operating in *ground effect*. Advance the throttle to where you did your last high speed taxi run and the aircraft is light on its feet — just below stall speed. Now, advance the throttle ever so slightly and, if the aircraft is trimmed properly, you should unstick. (A very slight touch of back stick may be required for a true unstick.) Hold it for an instant, (you should be just barely off the ground) then chop the power. You then should gently touchdown and roll to a stop. During that roll, you'll have to apply everything you've learned previously in controlling the post touchdown high speed taxi. Don't forget, you'll be rolling somewhere around 25 mph in a machine that's designed to fly, and not necessarily designed to be "driven." It might seem frightening at first, but practice will alleviate your fears as you become more familiar with the procedure. Again, practice makes perfect. And remember, always fly into the wind.

Once you've accomplished unsticking and touching down, you'll want to stay unstuck for longer than just an instant. Simply maintain the throttle at the setting that just produced a liftoff (full throttle is not needed nor recommended), for longer and longer times. As you progress, you may find the aircraft getting higher. It is prudent to decrease the throttle, and let the aircraft settle. You can glance at the wheels if you need to, but it is better to start using your perspective to judge height — it just takes practice. Remember, perspective is of utmost importance to takeoff and landing, especially landing. Besides maintaining proper airspeed, perspective is what landings are all about.

The power of your perspective vision is easily demonstrated. Stand at the end of a landing field, noting how things appear to your front and side. Now, get down on your haunches and notice how your perspective changes. Everything takes on a slightly different angle — different enough for you to tell whether you are rising, falling or at the same height! This is most significant, and is really what setting up a landing, approaching the runway and flaring are all about. Knowing your perspectives will enable you to make good, accurate landings.

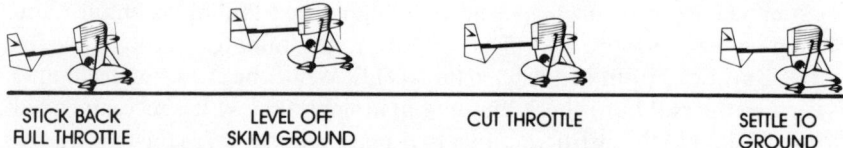

| STICK BACK | LEVEL OFF | CUT THROTTLE | SETTLE TO |
| FULL THROTTLE | SKIM GROUND | | GROUND |

Fig. 4-1. One possible method of crow hopping not discussed in text.

In the above description, we've assumed that the aircraft was properly trimmed, which your instructor can determine. Properly trimmed, your aircraft should unstick and skim the ground virtually without any control stick inputs. *Remember, the throttle setting determines whether you go up or down.* If, after touching down from a ground skim you find the aircraft bouncing, don't try to fight it. If you do, your control inputs will probably aggravate the bounces to the point where the plane begins acting like a bucking bronco. It is probably best not to move the stick at all, because the bouncing will stop as soon as the airspeed decreases. Make certain your aircraft is properly trimmed before you practice any crow hops. Let your instructor check it out first.

The Flare

As you continue your crow hopping, you'll want to advance the throttle a tad to increase your height. When you can successfully skim the ground for several hundred feet at a foot or so, you'll want to fly at, say, three to five feet. This should give you enough altitude to practice the flare. This maneuver is a part of many landings, and involves a gentle pitching up of the nose just before touchdown. It works something like this.

Let's say you've skimmed the runway at three feet for 500 feet, and you want to set down. Instead of just chopping throttle, which will allow you to gradually settle, chop the throttle and gently, very slowly, come back on the stick. Come back at a rate that gives you the impression or feeling that, if you keep coming back, the wheels will never touch. They, of course, will touch, but this technique should produce a smooth, bounceless landing. That's how my instructor taught me how to do it, and it works. Come back on the stick as if you were trying to hold the airplane off — and that's what a flare is. It will produce the so-called "three-point-landing" in a taildragger. Tri-gears will touch down on the mains first, the nosewheel dropping slowly as airspeed bleeds off.

Perspective is also important in the flare — you need to be able to judge whether you're sinking or climbing. You don't want to come back on the stick so quickly that you cause the aircraft to "balloon" and stall out. On the other hand, you don't want to sink too rapidly either, or you'll probably bounce. (Ultralights normally don't have shock absorbing landing gear, so bouncing can occur more easily than in a light plane). You simply must practice until you develop the proper judgement in determining your sink rate and back stick rate. Crow hops are great for developing the proper judgements.

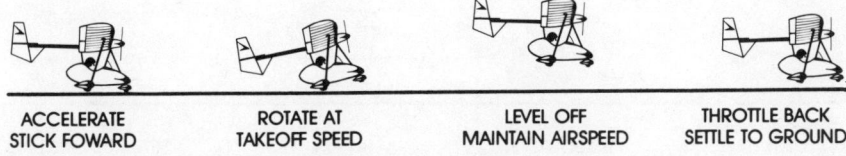

ACCELERATE STICK FOWARD   ROTATE AT TAKEOFF SPEED   LEVEL OFF MAINTAIN AIRSPEED   THROTTLE BACK SETTLE TO GROUND

Fig. 4-2. Crow hop technique for tri-geared ultralights.

Continue practicing your crow hops, gradually increasing throttle and altitude. As you gain height, remember to *always maintain flying speed — with the stick.* Forward stick lowers the nose and increases the airspeed. Back stick raises the nose and the airspeed decays. Avoid a stall at all cost — get that nose down if the airspeed gets near stall. In fact, you should be crow hopping at least 20% above stall speed. If your normal stall speed is 25 mph, you'll want to crow hop at an airspeed of at least 30 mph. Also, remember to strive for smooth control inputs. You're not riding a wild stallion, and jerkiness in control movements is uncalled for. If you're near a stall, lower the nose alright but, don't put the aircraft into a dive. The object is to maintain flying speed until you touch down. Strive for a smooth transition from flying to touching down.

Wheel Landings and Touching Down "Hot"

Nothing is more beautiful than a nicely executed flare to landing. They are fine whenever conditions are calm, or when a light, steady breeze is blowing straight down the runway. But, when things get gusty, or a bit "squirrelly," as ultralight pilots are wont to say, a flared type landing might not be the thing to do. Since the flare is done just a bit above stall speed, it doesn't take a genius to realize that a sudden gust, or change in wind direction, could easily stall the wing. This would immediately make your nose drop and, if you're near enough to the ground, you'll crash as sure as the sun rises in the east. Now, of course, you shouldn't be flying in gusty conditions to begin with but, if they arise during your flight, you'll have to contend with them. One method of minimizing having a gust induced approach to landing stall is to come in "hot," i.e., approach and touch down with extra airspeed.

Whenever the author is caught in a gusty landing situation, he likes to stay about 15-20 mph above stall speed. Keep the nose down and carry appropriate throttle to check the rate of descent. As you approach the ground, level out but keep some power on and "drive" the airplane onto the ground. Taildraggers would touch on main wheels only, holding forward stick pressure to keep them firmly planted on the ground. By the time your airspeed diminishes to the point where you can no longer keep the tail off the ground, you should be past the danger of a gust ruining your day. Once on the ground, you must still maintain vigilance, especially while turning away from the winds.

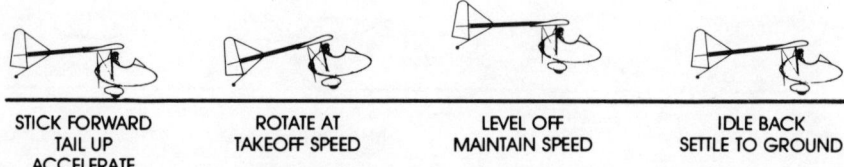

STICK FORWARD / TAIL UP / ACCELERATE — ROTATE AT TAKEOFF SPEED — LEVEL OFF / MAINTAIN SPEED — IDLE BACK / SETTLE TO GROUND

Fig. 4-3. Crow hop technique for taildragger ultralights.

Tri-geared ultralights can also handle gusty landing situations by carrying extra airspeed. The difference is in the touchdown — you won't be able to land quite as nose low as a taildragger, however, you can still use the basic idea. Simply land hot, with as low an angle of attack as possible, being careful not to touch on the nosewheel first — that can damage an otherwise perfect aircraft or hurt you. If you can touchdown "hot" on all three wheels simultaneously, that's fine, but definitely get the nosewheel on terra firma as soon as possible after the mains touch. Again, keep the stick forward, to plant the aircraft firmly on the ground.

Before you can hope to actually fly out of ground effect, you must become proficient at your crow hops. You should be able to fly a distance of a thousand feet or more at heights gradually increasing up to 50 feet or more. And, the emphasis here is placed on *gradually*. Increase the height of your hops gradually as you develop perspective judgement. Look all around you during these hops. You'll be better able to develop your sense of speed, and the way things appear from varying heights. This is essential practice, the importance of which cannot be over emphasized. You must be a master of crow hopping before you can really fly. See the chapter on landing for a more in-depth discussion and diagrams.

What About Turns

So far, no mention has been made of turning. That's because crow hops don't really take you high enough to turn in safety. However, if you find yourself drifting off the center of your intended crow hop path, you may need to apply corrective control. Remember, your crow hops should be done under calm conditions or straight into a very gentle (under 5 mph) breeze. There shouldn't be any need to turn, but that doesn't say there won't be.

If, while crow hopping, you find it necessary to turn slightly, it can be done with very gentle side stick pressure, toward the direction you want to go. That should be all that's necessary. To execute a proper turn requires a more detailed explanation than this, which we'll get to in a future chapter.

"Rotations about the longitudinal axis are governed primarily by the ailerons (or spoilers) which, when deflected, establish a rate of roll."

Takeoff and Climbout

5
Takeoff and Climbout

The crow hops you've been practicing and are now accomplished at, should have given you the judgement necessary to perform a real takeoff. In fact, you have actually been practicing takeoffs of sorts all along. The difference between a crow hop take off and a real one, will be your use of the throttle. Remember, crow hops are done in ground effect where less power is required to fly. Also, you are close to the ground and able to relate to your airspeed and height from your more familiar ground transportation experience. However, when you actually take off to really fly, the relationships of height and speed to the ground change.

A true takeoff, done with the purpose of leaving ground effect, requires full throttle. Compared to a lightplane, an ultralight accelerates to takeoff speed quickly and in a very short distance — some in around 50 feet. Furthermore, the angle of climb can be quite spectacular and a bit disconcerting to the lightplane pilot. With ultralight climb rates at from 500 to 1000 fpm or more, it isn't any wonder. Those rates, coupled with the fact that they occur at speeds in the low thirty mph range is the reason. You'll seemingly be "going-up-like-an-elevator" but should not be concerned, as long as you maintain proper climbout airspeed. This is one aspect of ultralight flying that licensed pilots are surprised and even frightened of but it's easy to get used to.

If you are in an open-air type ultralight, you'll have the additional potential problem of rediscovering the fear of heights. This is natural, but will soon disappear after a few trips around the pattern. The exhiliration of this type of flying is so great, that any fears you may have soon fade away into insignificance. If you are in an enclosed cockpit or have a fairing around you, this "fear" should not be encountered.

Some pilots might argue that the open-air, sit up front types offer no reference for judging attitude, but that's not necessarily so. You could and should arrange to mount a yaw string at the nose, and it'll give you an exact indication of the airflow angle relative to the airplane, which is far more important than attitude. Furthermore, attitude is easily judged by looking out across a wing and observing the angle it makes with the horizon.

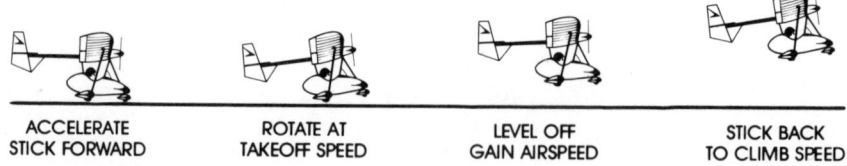

ACCELERATE STICK FORWARD   ROTATE AT TAKEOFF SPEED   LEVEL OFF GAIN AIRSPEED   STICK BACK TO CLIMB SPEED

Fig. 5-1. The sequence of a normal takeoff.

The Takeoff

As always, takeoff into the wind, i.e., head your ultralight in the direction from which the wind is coming. The purpose, of course, in doing this is to benefit from the "free" airspeed you gain from the wind, enabling a liftoff at less of a ground speed, and therefore less of a ground run. Taking off into the wind will allow you to climb higher in less ground distance covered, adding to the safety of the event. Bear in mind however, that you should be doing your initial familiarization under calm conditions, or only in gentle breezes not exceeding 5 mph.

Crosswind takeoffs are definitely not recommended in the beginning of your ultralight experience, but may be attempted later on after you become more proficient. The better three axis control ultralights can handle a moderate amount of crosswind, but leave it for later. Crosswind operations will be discussed in an upcoming chapter.

Theoretically, if your ultralight was pointed into a wind equal to its takeoff speed and you "poured-on-the-coal", it would rise straight up. But only a fool would try such an experiment. Unless the wind were perfectly steady, smooth and head-on, an ultralight would not only be lifted up, but also be blown away, out of control.

If conditions are even mildly gusty, you could be asking for trouble during a takeoff. I've seen ultralights lifted off before full flying speed was reached,

Takeoff and Climbout 47

only to be unceremoniously and abruptly slapped back down onto the runway. These conditions are not recommended, although, keeping the ultralight on the ground by holding forward stick, while you accelerate to well above your normal takeoff speed, could get you through a sudden lull or gust. Personally, I always carry extra airspeed not only in the takeoff, but during the landing as well. It serves as extra insurance against being "swatted" out of the air.

Takeoff Procedure

The basic procedure in making a normal ultralight takeoff, is as follows (after a normal preflight, of course):

1. Warm the engine. Let the cylinder head temperature come up to the recommended value and be able to advance to full power and back off without missing or backfire. If you have dual ignition, check both circuits — as indicated by a drop in rpm when one is shorted out — as on any lightplane. You should, of course, be doing this engine "run-up" check while facing into the wind.

2. "Clear" the area — be sure there are no other aircraft, ultralight or otherwise, in the pattern. If you can't see all around the pattern from your seat, do a 360 taxi to see what's going on.

3. If all is clear, double check the wind direction and taxi to the runway. Remember how to handle crosswinds during your taxi, they can be quite "upsetting" if ignored.

4. Line up into the wind at the end of the runway, and open the throttle wide.

5. Hold the stick forward and accelerate to a little more than your best angle of climb speed (approximately 1.2 times stall speed).

6. Gently ease back on the stick, rotate and lift off.

7. Continue accelerating in ground effect to slightly above your best rate of climb speed (about 1.4 times stall speed).

8. Next, ease back on the stick to your best rate of climb speed and climb until a few feet below the desired altitude.

9. Lower the nose to cruising speed, then throttle back to cruise rpm.

A licensed private pilot could master the ultralight takeoff in a few tries. It all depends on his proficiency. A person who has never flown could possibly do as well, or maybe even better, but probably not. It's hard to say exactly how many practice takeoffs it might take for a novice to develop the necessary skills. Perhaps as little as 20 or 40. Your flight instructor will tell you when you're ready. And, dual instruction is definitely recommended.

The Climbout

The climbout is, of course, the natural extension of the takeoff run. And once again, as in all phases of flight, it is necessary to maintain proper airspeed, or more precisely, angle of attack. Rather than "chasing-after-the-airspeed-indicator" though, concentrate on the aircraft's attitude, and

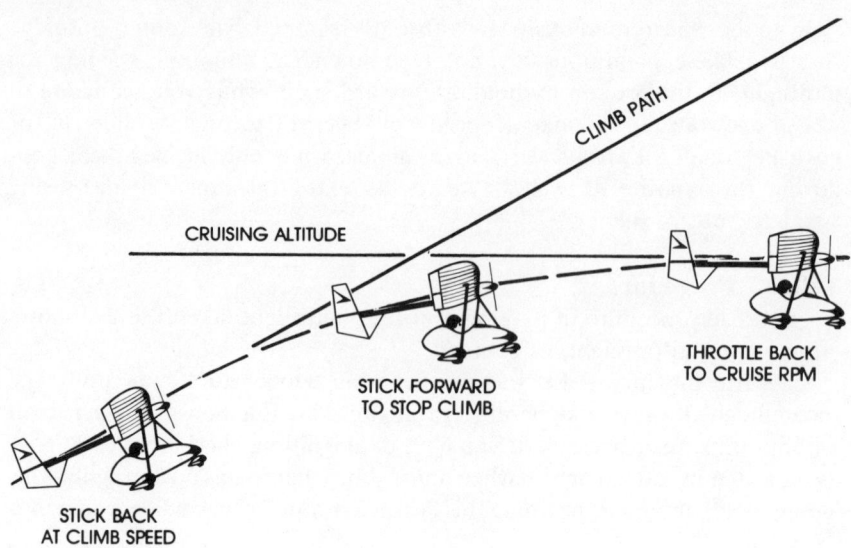

Fig. 5-2. The climbout is completed just below the desired cruising altitude.

takeoffs should become easy. Once your airspeed (stick position) and throttle (rate of climb) are set, virtually no corrections should be necessary after the climb is established. Only small stick movements, on the order of a half inch or less, might be needed, depending on the roughness of the air.

When you are anywhere between 3 and 5 feet below your desired cruising altitude, the nose should be lowered to slow the altimeter (rate of climb) in anticipation of leveling off. When the desired altitude is reached, the throttle can be adjusted to cruise rpm to maintain cruising speed. If your ultralight is properly trimmed, this is where it will show up — no stick forces should exist during cruise. If you have to apply pressure to the stick at your desired cruising speed, the airplane will need retrimmed.

NOTE: If a prolonged slow climb is required, to clear obstacles or whatever, keep a close watch on cylinder head temperature. You wouldn't want the engine to overheat and seize. If the head temperature is climbing, lower the nose and reduce power a bit until it cools down.

Clearing Obstructions

To get out of really confined fields, we must use the best angle of climb speed, which is normally about 1.2 times stall speed. Accelerate on the ground to slightly less than the best angle of climb speed, then rotate, lift off, and climb out at the best angle of climb speed.

Extra caution must be exercised during a best angle of climb climbout, especially during gusty conditions. Here, your airspeed is only 20% above stall, and a gust could easily stall you out. Furthermore, if you have an engine failure during the climb, you'll have only an instant to get the nose down to

Takeoff and Climbout

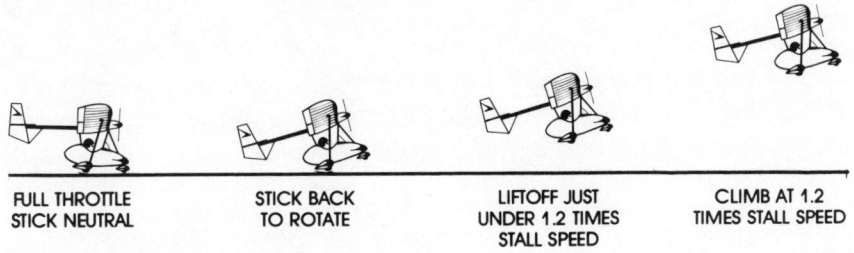

FULL THROTTLE	STICK BACK	LIFTOFF JUST	CLIMB AT 1.2
STICK NEUTRAL	TO ROTATE	UNDER 1.2 TIMES	TIMES STALL SPEED
		STALL SPEED	

Fig. 5-3. The short field takeoff procedure.

accelerate to your best gliding speed of 1.4 times stall. This puts you in a precarious situation, so best angle of climb climbouts are not recommended. Use best rate of climb (1.4 times stall speed) climbouts for a better safety margin, in the event of an engine failure.

Soft Field Takeoff

When field conditions are a little on the soggy side, you'll want to lift off as quickly as possible. The object is to unweight the wheels as soon as you can in the takeoff roll. The basic technique involves full throttle, back stick and flaps, if you have them. Accelerate, and lift off at or slightly below the power-off stall speed — you're being aided by ground effect. As soon as you lift off, gradually reduce your pitch attitude until you are skimming the ground in level flight. This will alow you to accelerate to your normal climb speed and climbout.

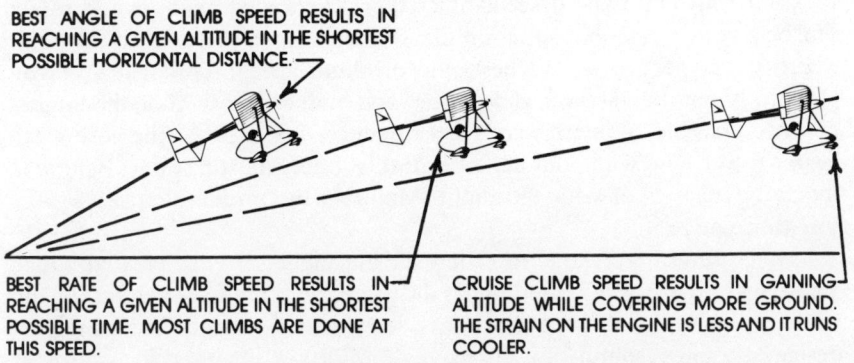

BEST ANGLE OF CLIMB SPEED RESULTS IN REACHING A GIVEN ALTITUDE IN THE SHORTEST POSSIBLE HORIZONTAL DISTANCE.

BEST RATE OF CLIMB SPEED RESULTS IN REACHING A GIVEN ALTITUDE IN THE SHORTEST POSSIBLE TIME. MOST CLIMBS ARE DONE AT THIS SPEED.

CRUISE CLIMB SPEED RESULTS IN GAINING ALTITUDE WHILE COVERING MORE GROUND. THE STRAIN ON THE ENGINE IS LESS AND IT RUNS COOLER.

Fig. 5-4. How various airspeeds affect climb performance.

Engine Failure on Takeoff

If your engine seizes or otherwise fails during the takeoff, the first thing you must do is to lower the nose to obtain your best gliding speed, which is just about the same as your best climb speed, and keep cool. If you are high enough to turn back to the field, fine, but maintain that airspeeed or you'll

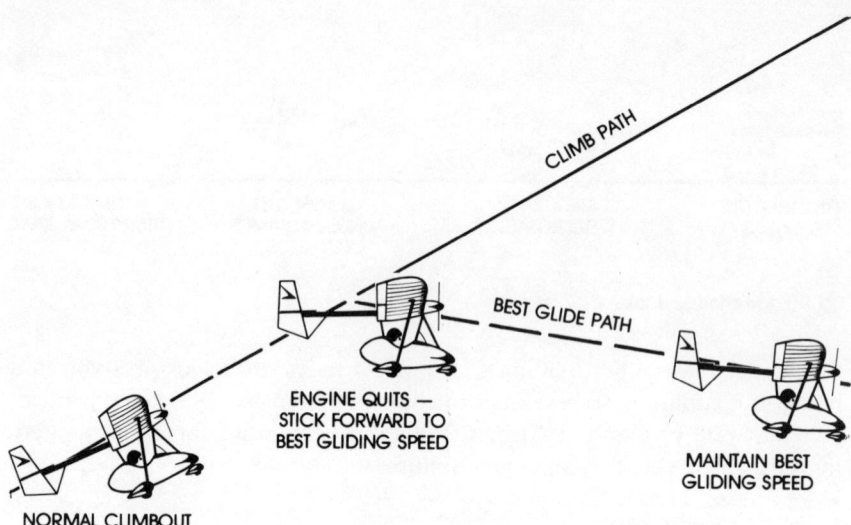

Fig. 5-5. Procedure for handling an engine failure during takeoff and climbout.

stall and possibly enter a spin. (The stall/spin is the major cause of flying accidents and deaths!) If you are too low to make a safe turn, it's probably best if you continue straight ahead or initiate a very shallow bank to avoid hitting obstructions. Do not attempt a 180° turn and downwind landing unless you are absolutely certain you have enough altitude to do so.

Whatever you do, don't try to "stretch the glide" by pulling back on the stick. It'll only make you descend at a steeper angle and possibly cause a stall. The best you can do in a situation like this is to lower the nose to the point where your speed is at the best rate of climb speed. This will give you practically your shallowest glide angle, and enable you to glide the longest distance possible. While it may seem disconcerting to lower the nose when nearer the ground with your engine out, this simply must be done. Believe in the aerodynamics of your ultralight. Maintain the proper airspeed for the situation you're in.

It is important to remember that all ultralights are not created equal. Those with wire bracing offer higher drag and therefore less penetration and probably slower stall recovery. The cleaner strut braced and cantilever designs should respond quicker to angle of attack changes, in regards to picking up airspeed, and aid stall recovery. Higher drag designs also offer steeper glide angles than cleaner ones. Get to know the gliding characteristics of your ultralight before you are forced to use them.

6
Bank and Turn

Once you have successfully mastered the takeoff and find yourself cruising, you are faced with the task of either maintaining a direction of flight and/or turning. To gain a clearer picture of what is involved in controlling flight direction, let's use the example of a car traveling on a slippery road.

Suppose you are moving along on a straight and level highway at a fairly high rate of speed, as if the road were dry, and you attempt to negotiate a curve. Not long after you enter the curve, you find yourself skidding sideways off the road. If the road had been "banked" enough though it should be apparent that the skid would not have occurred.

This situation is quite similar to flying a turn. For instance, if only a rudder is deflected, say right, while the ailerons are not touched, the nose will swing toward the right, but the airplane will not come around, without skidding toward the original direction of flight. The airplane wants to keep flying in the direction it was going, because of its inertia. To correct the skid, the wings (like the road) need to be banked to tilt the lift force toward the center of the turn to counteract the centrifugal force tending to make the aircraft not want to turn.

Now, suppose you tried to drive your car along the bank of a slippery hill. You couldn't! The car would, of course, slip sideways until you reached the level (un-banked) bottom. No amount of turn away from the fall line would prevent the slip — only the leveling of the surface. Alternately, the car could be driven down, if the front wheels were steered toward the bottom or level area.

This situation is also similar to flying. If an airplane is banked only, with insufficient rudder into the turn, it will slip toward the low wing toward the center of the turn. If the rudder is deflected into the turn, a coordinated turn will be established and the slipping will stop. Alternately, the wings could be un-banked, i.e., leveled, accomplishing the effect of the car reaching the level at the bottom of the hill, and the slip would be stopped!

Banking is What Turns an Airplane

If you were to drive your car on a properly banked, icy racetrack at some given speed, depending on the degree of banking and radius of curvature, you would neither skid nor slip — you'd be perfectly balanced or coordinated in the turn. If the slope of the bank were too steep for your speed and turn radius, you'd slip toward the center of the track. If the track wasn't banked steeply enough, you'd skid to the outside and hit the guard wall!

This situation is exactly analogous to flying, but in flying it's the bank angle that establishes the turn rate for a given speed. When a wing is banked, it tilts its lift toward the center of the turn, in essence, pulling the aircraft around the turn. The amount of pull (lift) is, of course, dependent upon the angle of attack, or pitch attitude relative to the airflow (not with respect to the horizon!).

While an aircraft is banked and its lift is tilted toward the center of the turn, it still needs to develop a vertical component of that lift equal to the gross weight, to keep itself up. What this means, is that unless the stick is pulled back and/or the throttle advanced, the aircraft will descend. Now remember that stall speed increases with bank angle, so it won't take as much back stick to stall, while banked. Stay above the stall speed for your angle of bank and you won't stall!

To increase the turn rate, it is necessary to increase both the bank angle and the angle of attack. The bank angle will determine the turn rate for a

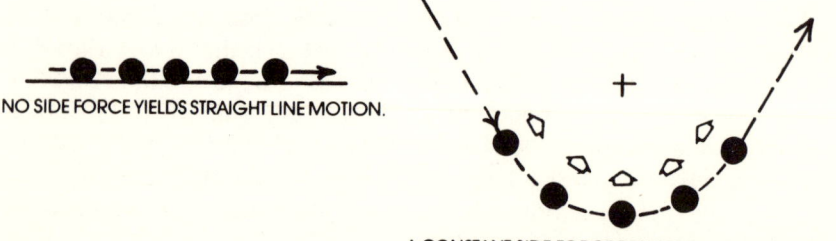

Fig. 6-1. A constant side force is needed to produce a turn.

Bank and Turn

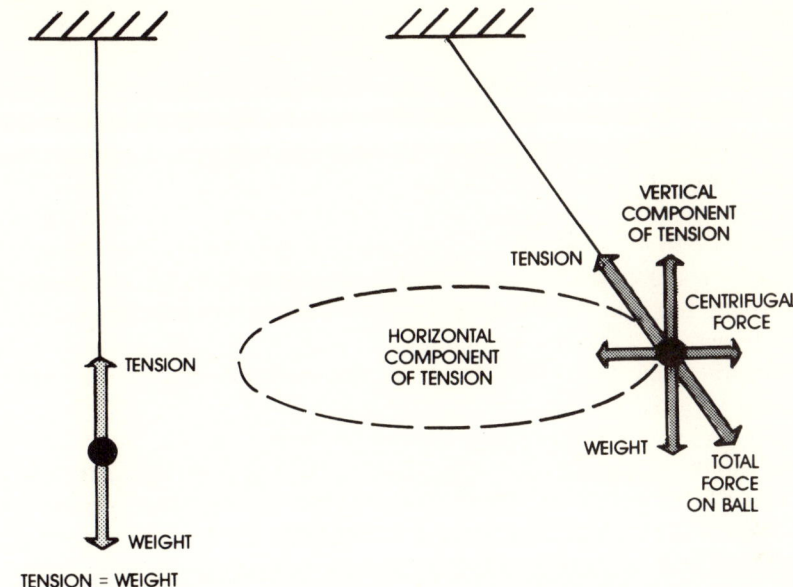

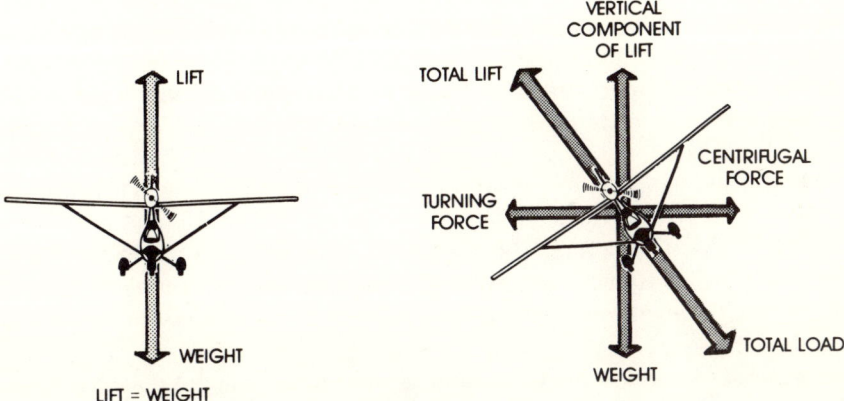

Fig. 6-2. Understanding the physics of a turn.

given airspeed, while back pressure (angle of attack) will allow generation of enough vertical component of total lift to maintain altitude. The turn rate can be decreased by shallowing the bank and reducing the back pressure.

As a general rule, it is a good habit to enter a turn with more speed than when you were flying straight and level. This may be done by advancing the throttle and/or pushing the stick forward. This will not only help get you away from the banked stall speed, but also make the controls a bit more responsive. Once in the turn remember that the THROTTLE CAN BE USED FOR AIRSPEED CONTROL TO TIGHTEN THE TURN

WHILE HOLDING BACK PRESSURE, OR TO CLIMB AND DESCEND.

Throughout the foregoing discussion, you may have noticed that no mention was made of the rudder! What is its function in a turn? Simply this. The rudder is used merely to counteract any yaw tendencies that develop while the other controls are used — that is all!

But what about the so-called two control (rudder and elevator) ultralights. After all, they don't have ailerons, yet they are turned by rudder, you say. Well, to the casual observer, it certainly looks like the rudder does the turning, but it isn't so. This apparent paradox can be explained as follows.

Turns can certainly be made by using the rudder, provided enough dihedral (wide "V" formed by the wing panels) exists. When the rudder is deflected, it first causes the aircraft to yaw, and since it doesn't have any ailerons to set up the appropriate angle of bank, the aircraft enters a skid. Due to the dihedral angle, the underside of the outside wing is exposed to the skid velocity which in turn lifts that wing into a bank. The wing lift is tilted toward the center of the turn and the airplane turns! Now the elevator and throttle can be used as described above to control the turn. Properly designed, the aileronless ultralight can actually make fairly coordinated turns. However, an airplane without ailerons is quite limited in the amount of crosswind it can handle, and in its ability to be rolled. A wing simply cannot be "picked-up" right now without ailerons to roll the airplane.

Judging the Proper Angle of Bank — Coordinating the Turn

When the angle of bank is too steep for the airspeed and radius of curvature desired, the aircraft will slip toward the inside of the turn and the pilot will feel himself sliding sideways toward the low wing. This can, of course, be corrected by decreasing the bank.

If the angle of bank is too shallow for the airspeed and radius of curvature desired, the aircraft will skid toward the outside of the turn and the pilot will feel himself being thrown toward the outside of the turn. The solution would be to increase the angle of bank.

Without the aid of a bank indicator, which most ultralights don't have, the proper bank angle can be felt by the seat-of-the-pants. A perfectly coordinated turn will give the pilot no side force sensations. He'll just feel as though he were in straight and level flight, while observing an apparently titled horizon moving by sideways in front of him. He will be neither slipping or skidding.

How a Turn is Made

In making a coordinated turn then, an airplane must be rolled, yawed and pitched in a smooth continuous motion. For the beginner, however, it is possible to perform this in steps. For example, to make a gentle turn of moderate radius to the left, start by rolling the aircraft into a moderate left bank. Push the stick to the left until the desired bank angle is reached, then

Bank and Turn

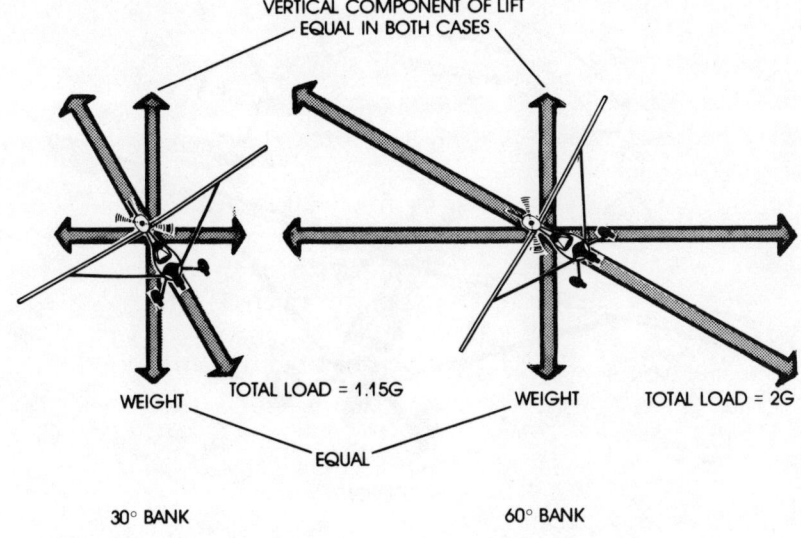

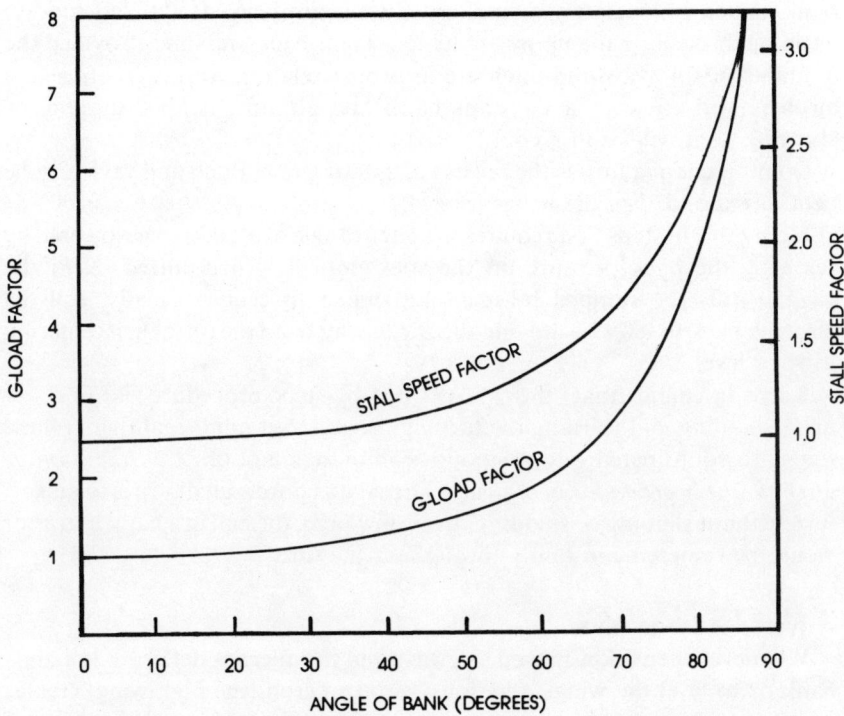

Fig. 6-3. G-loads and stall speeds increase with bank angle.

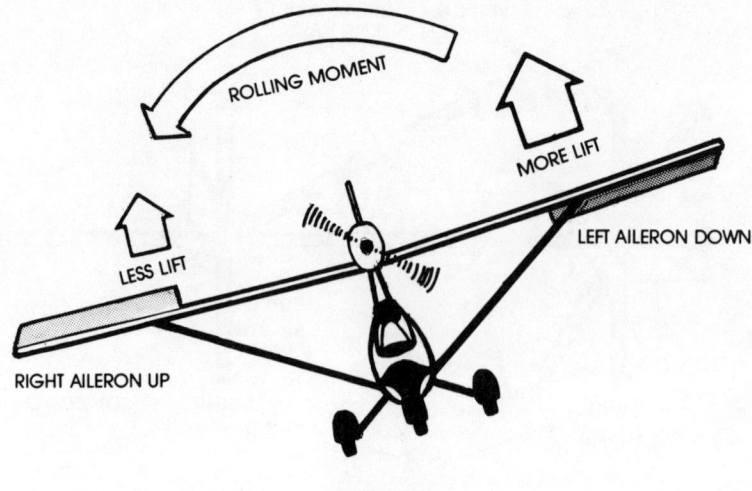

Fig. 6-4. Aileron deflection developes unequal lifts on each wing panel, causing an air vehicle to roll.

move the stick back to neutral (or slightly to the right, as required to hold the bank angle). Immediately thereafter, yaw the airplane left with left rudder. Then pitch the airplane up by applying a slight back pressure. Provided the amounts of roll, yaw and pitch are in proper relation with each other, the airplane will circle at a constant bank and altitude, neither slipping or skidding — it will be in a coordinated turn.

Coming out of a turn is the reverse of entering one. Pitch and yaw must be neutralized and *then* the wings leveled.

Doing this in steps first requires the pitch (angle of attack) to be lowered by releasing the back pressure on the stick until it is neutralized. Next the yawing must be stopped by releasing rudder to center. Finally, roll the airplane back to level by moving the stick away from the turn, then centering it when level.

Keep in mind that the above step-by-step procedure is fine for understanding and learning the turning process, but in normal flying, these steps are coordinated into a smooth continuous motion. *Coordination is, after all, the heart and soul of flying.* To make a coordinated turn is to make a turn without slipping or skidding. In other words, the ball in a bank indicator would be centered constantly throughout the turn.

Adverse Aileron Yaw

Whenever the stick is moved sideways and the ailerons deflect to initiate a bank or to level the wings, the down going aileron (the high wing) creates more drag than the up going (the low wing) aileron. This has the effect of actually yawing the aircraft in the direction opposite to the intended turn, especially at slower speeds (high angles of attack)! Because of this effect,

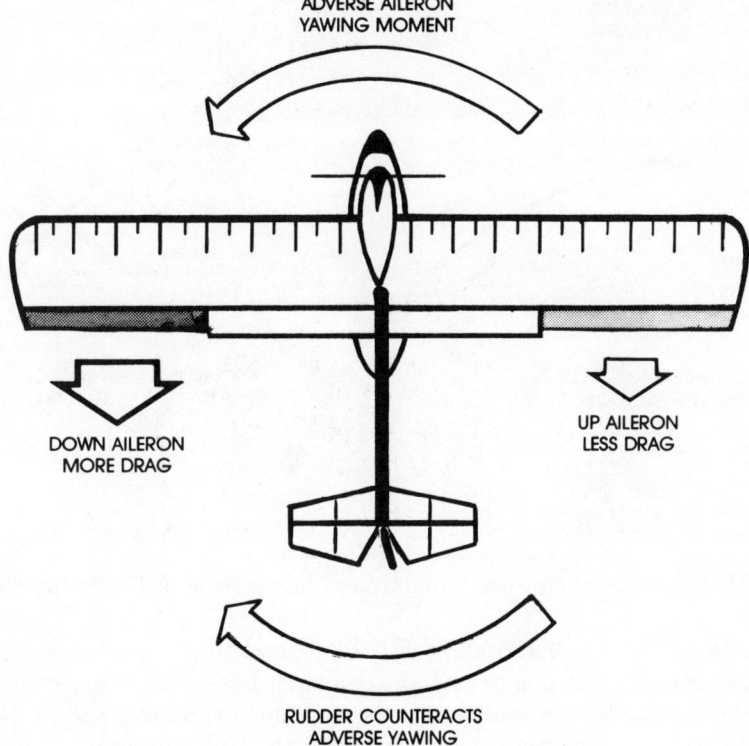

Fig. 6-5. Adverse aileron yaw causes an air vehicle to yaw in the direction opposite the intended turn. Proper use of rudder will compensate for it, and coordinate the turn.

called adverse aileron yaw, it is necessary to apply rudder into the turn, otherwise, you will be in an uncoordinated turn — slipping slightly.

The Rudder is Not Necessary To Turn

If an ultralight is well designed, and possesses a certain amount of weather cocking tendency, it is possible to make a coordinated turn without the rudder deflected. If there is sufficient side area aft the CG, the aircraft will tend to position itself with the circumference of the turn so that no rudder is needed. This is especially true of higher speeds where minimal aileron deflection is required to bank, thereby creating little or no adverse aileron yaw.

Understanding the Physics of a Turn

In high school we learned Newton's first law of motion: "A body persists in its state of rest or of uniform motion in a straight line unless it is compelled to change that state by forces impressed on it." And so it is with an airplane flying in straight and level, unaccelerated flight. Unless a force is introduced

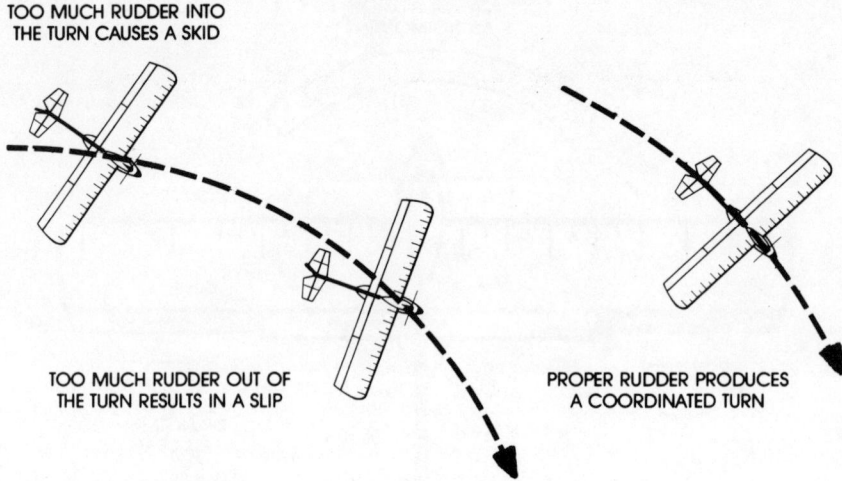

Fig. 6-6. How rudder is used to coordinate a turn.

to unbalance the established equilibrium, the airplane will continue doing what it is doing.

To gain a better understanding of the forces generated in a turn, we can use the example of a bowling ball. Once it is released down the alley, it will roll straight — the alley is smooth and there is nothing to interfere with the ball's trip to the pins. The only way we can alter the ball's course is to apply a sideways force. And, the only way we can make it turn in a circle is to apply that side force continuously, both perpendicular to the direction of motion and toward a single point. This "toward-a-single-point-side-force" is known as centripetal (which means center seeking) force, and it counter-balances the centrifugal force, which tends to tear the ball away from the center.

Now let's take our bowling ball and suspend it from a rope. The tension in the rope is upward and opposite the ball's weight — a situation analogous to straight and level, constant speed flight where lift equals gross weight.

Unlike the bowling alley situation, the ball is now restrained. If we give it a push this time, it will move in a circle at some "bank angle" depending on its speed. Here, the rope's tension is comprised not only of the ball's weight, but also the centrifugal force! And, this is exactly as it is in a turning airplane, where the lift is equal to the tension. The lift must be greater in turning because it's not only counterbalancing the gross weight, but also the centrifugal force trying to "un-turn" the airplane. The inward pointing lift is what turns the airplane.

Pivotal Altitude

Pivotal altitude is that height above the ground, for a given speed, at which an airplane can fly around a point while its lateral axis is pointed at that point. This altitude is always the same, regardless of the turning radius. The

Bank and Turn

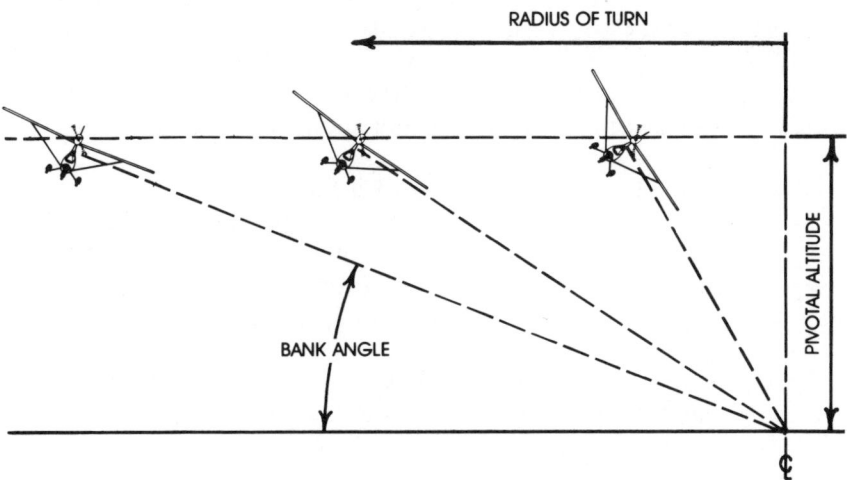

Fig. 6-7. There is a certain pivotal altitude for any given airspeed and angle of bank in a coordinated turn.

shorter the radius the steeper the required bank.

At 50 mph, for example, the pivotal altitude would be 168 feet. At 25 mph it would be 42 feet. Pivotal altitude varies as the square of the speed, i.e., doubling airspeed quadruples the pivotal altitude.

Now keep in mind, if there is a breeze, the angle of bank has to be continuously varied to turn about a point. With the wind coming from the inside of the turn, a steeper bank is needed to prevent drifting with the wind, and vice versa. The faster the groundspeed caused by the wind, the steeper the angle of bank required to maintain a constant radius. Only practice will make turns about a point possible in a wind.

Steep Turns

Sometime in your flying career, you may find it necessary to do a steep turn, and anything above a 30° bank might be considered steep. At 60°, you'll experience a 2-'g' load (pressed into your seat at twice your weight), the wing will have to develop twice its normal lift, and you'll have to maintain an airspeed of more than 1.41 times your normal stall speed. While 60° of bank might not sound like much, from an airplane it looks as though you are almost vertical!

The main thing about a steep turn is that is requires power levels above those used for normal cruise. Before entering the turn, advance the throttle and build up excess speed, at least 1½ times (for a 60° bank) your normal speed, but more to move well above your stall speed at 60°. Then enter the turn as you would any other, with the correct amounts of pitch and yaw and continue banking until you're at the desired angle.

The other thing about a steep turn is that, besides the bank angle itself, pitch control will be most important. Yawing required is minimal. If you pull too much back stick, you'll stall. If you pull too little, you'll start slipping into the turn and lose your altitude.

To stop a steep turn, do as with any other turn. Reduce the back pressure to lower the angle of attack, *then* roll out and correct for any yawing.

Turning Onto The Downwind Leg of a Pattern

In general aviation, there seems to be continual discussion about the dangers of the so-called downwind turn. And, indeed there should be, as it seems to be the focal point for a significant percentage of all flying accidents. So what extra precautions should the ultralight pilot take when turning downwind?

First of all, when flying at an altitude safe enough to recover from a stalled turn, there shouldn't be any more concern than normal: besides, a good pilot won't stall in a turn. When near the ground, however, it's an entirely different ballgame. But why should the propensity to stall be greater near the ground?

As we've said many times before, an airplane simply will not stall, provided it stays below stall angle of attack. That sounds easy enough, but in making a downwind turn near the ground a couple of "culprits" emerge. For

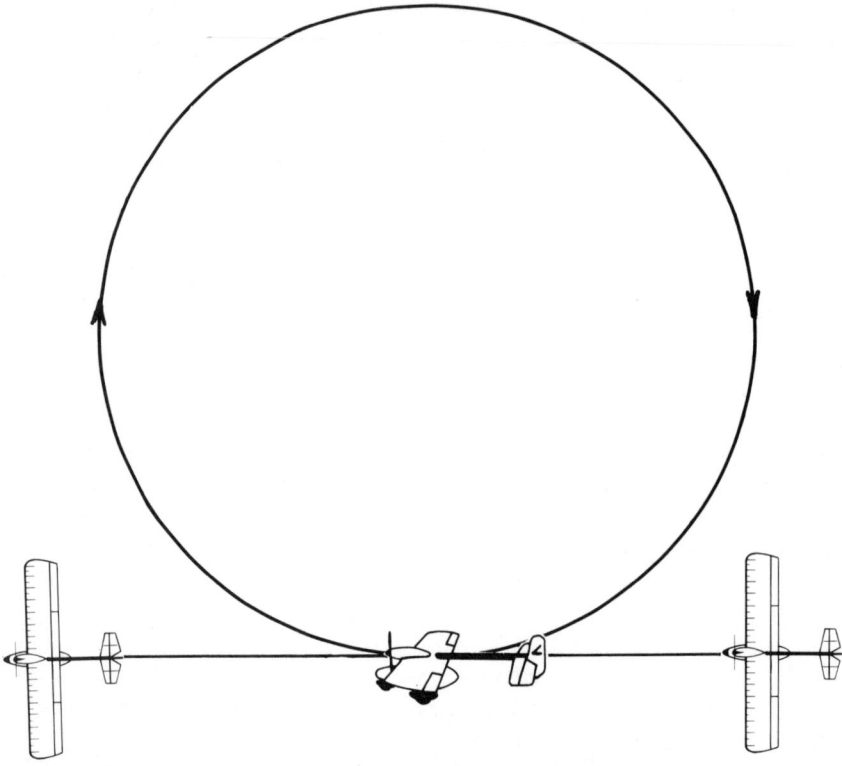

Fig. 6-8. A constant banked turn in a calm produces a circular ground track.

Bank and Turn

example, let's assume you're making a forced landing in the downwind direction because there is no other alternative. After determining your perspective, you set up a base leg, with a crosswind to your right. You begin turning left onto final, and here's where you make or break the landing.

In this situation, that crosswind could tend to over bank the aircraft. Also, the turn might be a little tighter than it should be (to make the field) and the stall speed at that bank angle could be exceeded. The air near the ground is often turbulent and a sudden gust could stall a wing. Then, too, once you've turned final, the tailwind will add to your ground speed. This gives the false impression of excess airspeed, causing you to incorrectly apply back pressure, resulting in an approach to landing stall.

To avoid the trauma and potential dangers associated with downwind turns, the rule is simple — MAINTAIN ANGLE OF ATTACK BELOW THE STALL. One way to do this is by keeping your airspeed well above the steep bank stalling speed of the airplane. It's far better to have too much airspeed than too little, especially in an ultralight. And, don't let ground speed fool you.

Clearing Turns

When you wish to make a turn in a high wing ultralight, it makes good sense to "clear-the-turn" first. This is done simply by first momentarily raising the wind on the inside of the intended turn and checking for traffic in that direction. If all's clear, proceed with the turn. Low wing ultralights don't have to clear-the-turn, since the inside wing clears the view as it drops.

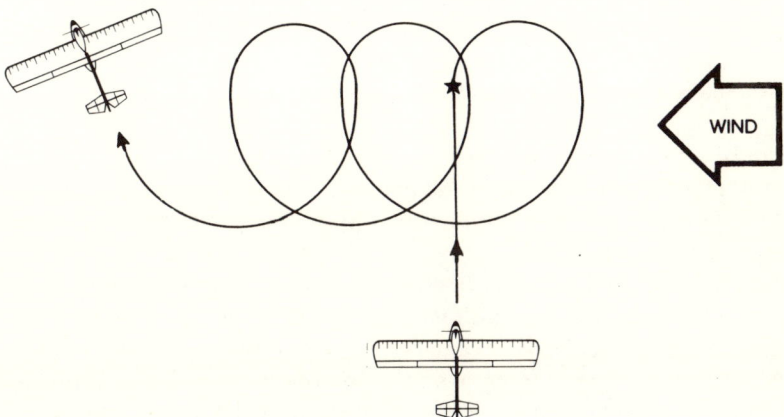

Fig. 6-9. How wind affects a constant banked turn.

Turning About a Point

When a constant banked turn is made under calm conditions, the ground track is a perfect circle. When a wind is blowing however, the airplane

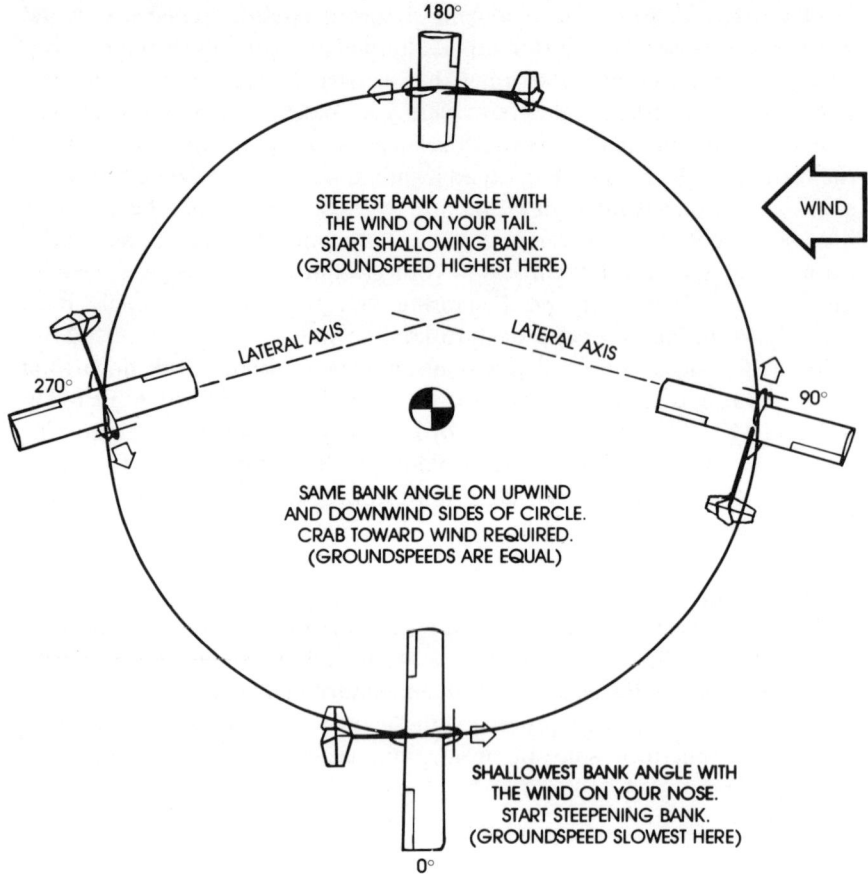

Fig. 6-10. How to do a constant radius turn in a wind.

naturally drifts with it, complicating the maneuver. A constant bank turn done in a wind will result in an oblong circle ground track.

To fly a constant radius turn about a point in a wind, the angle of bank must be varied continuously during the turn. Starting the turn into the wind, we would be at our shallowest bank angle. Citing the ground object (which may be a barn, for instance) with reference to the wing, we established a rate of roll, (holding aileron as necessary) increasing our bank. We must also continuously increase our crab angle, for the wind becomes more of a crosswind as we approach the 90° point of the turn. The crab angle will have our ground point drifting further back in our view as we approach the 90° point, as well.

Once at the 90° point in our turn, we must begin yawing *toward* the ground object, and finally have it in line with our aircraft reference point, at the 180° point of the turn. This point will also have us at the steepest bank angle, too.

From the 180° point on, it will be necessary to begin decreasing our angle of bank, while increasing our crab into the turn. The object we are circling around will begin to drift forward of our original aircraft reference until the 270° point in the turn. Here, our crab angle will be equal to what it was at the 90° point. From the 270° point to the original starting point, the crab will be decreasing, until it is zero, and the ground object once again lines up with our original aircraft reference point.

As with all the aspects of flying, the only way to do turns about a point is to practice. Have your instructor show you how to do it in a two-seat trainer, then go practice on your own. Be certain you are at a safe altitude and not over a congested area. And, always be sure to "clear" the area by checking for other traffic before you begin this or any maneuver. Having your instructor on the ground while circling and in radio contact, would be an excellent way to judge your proficiency at doing turns about a point.

"The essential element of flight is that the wing be unstalled."

7
Mushes, Stalls and Spins

The Mush

Many of the ultralights are often described as being able to mush. Some are also claimed to go into a mush, rather than a stall, when the stick is pulled all the way back.

So what is a mush? Somewhere between a normal glide and the actual stall is where you'll find it. Once in a glide, if you gradually come back on the stick and stay slightly above the stall, you'll be mushing. It's a glide with a higher sink rate and/or steeper descent path, while the nose (angle of attack) is held higher than for a normal glide.

In the case of a mush, the higher angle of attack causes more drag since you're gliding below your maximum glide ratio. Mushing can be used to steepen the approach, but it is considered dangerous since it is done so near the stall. And, if you were to land in a mushed attitude, you'd certainly damage the airplane and injure yourself, because of the sink rate.

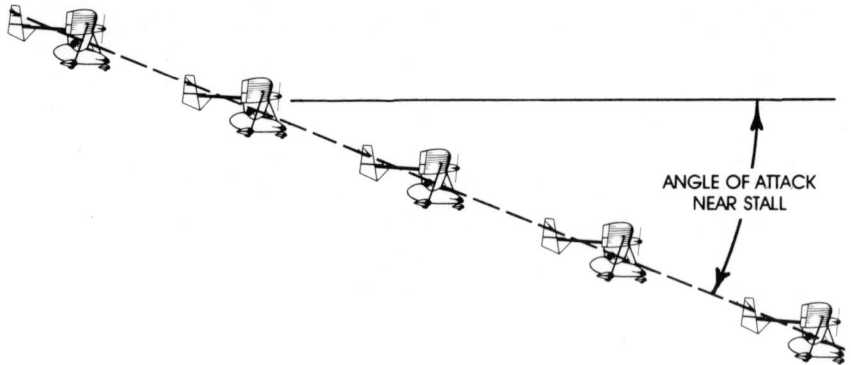

Fig. 7-1. Mushing is dangerous because its done so near the stall. This is especially true during gusty conditions.

Stalls

All throughout this book we've mentioned the stall, but let's look at it in more detail. Before we do, remember that any ultralight can be made to stall. Sure, some are more resistant to stalling and require a different technique, but they can all be stalled, and at any airspeed!

A stall is that event which takes place when the wing is at such a high angle of attack that the airflow over the upper surface breaks away, diminishing the lift. As a result, the nose drops and the aircraft dives until the angle of attack is low enough for the wing to develop its full lift. The proper angle of attack for airflow reattachment coincides with an airspeed just above the stall speed.

The stall can be recognized in ways other than by what the airspeed indicator says. Some ultralights might exhibit a buffet, and shaking of the stick, as the airflow partially separates then reattaches to the wing's upper surface. That's the time to get the nose down (lower the angle of attack) before the stall breaks.

If a stall is left to occur and the pilot does nothing, the nose will drop and the airplane will dive until it has enough speed to fly and then, it will gradually level itself out. If the pilot wants to, he can improve upon the airplane's natural recovery rate by gradually coming back on the stick after the nose has dropped, that way affecting a recovery with a very small loss of altitude — maybe within 25 feet!

Stall recovery is something every pilot must be capable of doing — his life could depend on it, especially if he's near the ground. Keep in mind that roughly half of all aviation accidents are stall related. Stalls should, of course, be practiced but only at a safe altitude, with plenty of room for recovery.

Stalls can be made with power-off and power-on. The power-off stall can be done with the engine at idle, the airplane in a glide. Gradually, pull back

Mushes, Stalls and Spins

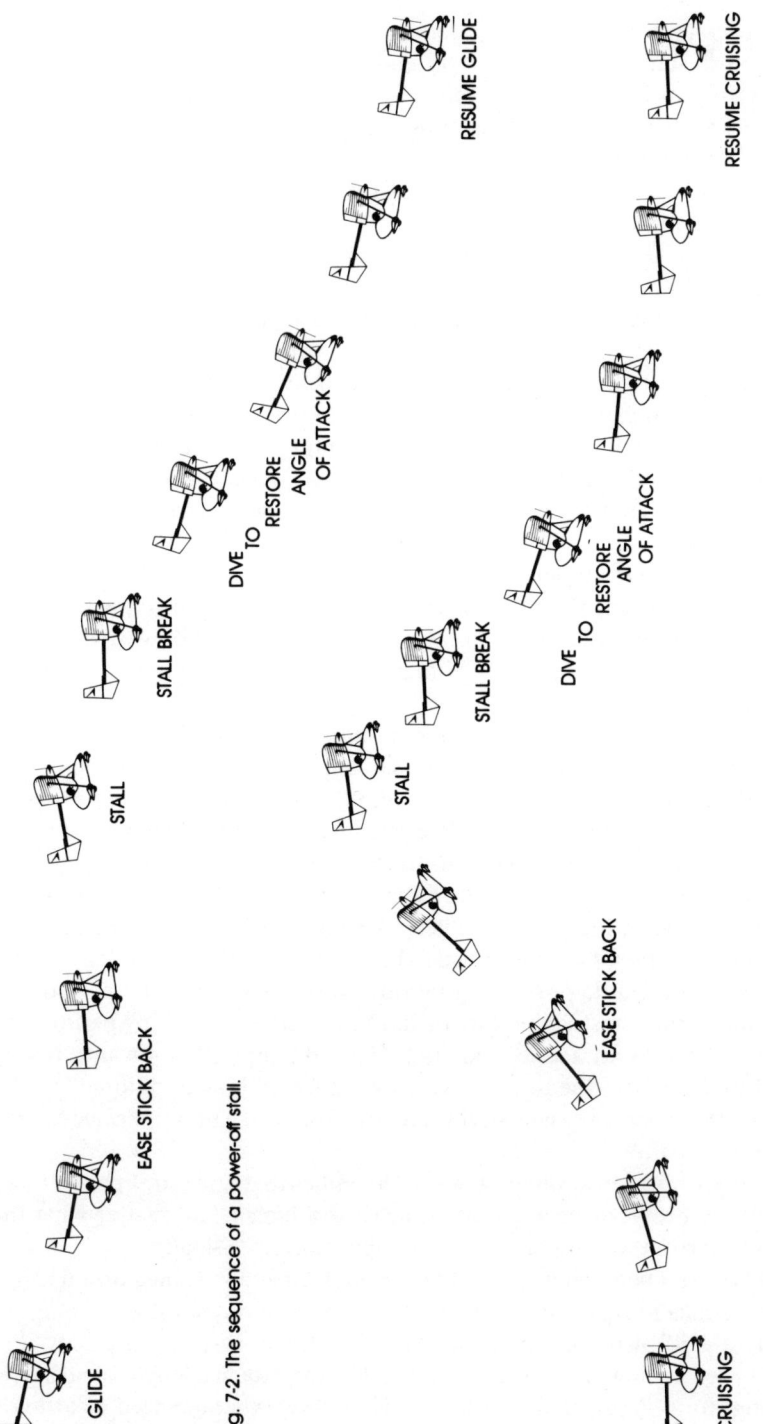

Fig. 7-2. The sequence of a power-off stall.

Fig. 7-3. The sequence of a power-on stall.

on the stick and try to hold the aircraft level. The airspeed will drop and the airplane will stall. As soon as you recognize the stall, push the stick forward, regain flying speed and resume your glide. (If a gradual application of back stick doesn't result in a stall, you might have to come back sharply on the stick to force a break.)

The power-on stall requires a gradual application of back stick as you put the airplane into a very steep attitude, say 30°-45°, with the throttle set at cruise. After a few seconds, the stall will occur and either the nose will fall through or the aircraft will sink. Recover by pushing the stick forward and return to level flight as soon as flying speed is regained. (Be careful not to over speed the engine during the possible ensuing dive).

While in a stall, keep in mind that the airflow is separated from the upper surface of the wing, making the ailerons useless. In fact, if you are near a stall and wish to raise a low wing, forget it. The instant you move the stick, the low wing will stall and possibly put you into a spin. The only thing you can do is to lower the angle of attck first, regain attached flow (flying speed) and then level the wings.

The rudder, on the other hand, is the least affected by the stall and it can be used to aid in keeping the wings level, as well as in maintaining direction. The rudder is especially effective in a power-on stall, with the propwash making it so. In a power-off stall, much rudder movement is required to maintain effective control.

At this point, it is necessary to mention the canard (small wing in front) ultralight and its supposed stall resistance. This configuration is set up so that a stall is just about impossible — just about. Under normal circumstances, the canards do seem relatively stall resistant! If the nose is brought up to the normal stall angle, what happens is that the canard (which might be supporting only 20% or less of the aircraft's gross weight) stalls, causing the nose to drop while the main wing keeps on lifting, and the airplane does not stall. In fact, if the stick is continued to be held back, the canard will rise, stall, then lower in a gentle porpoising motion, with a characteristic period.

The apparently unstallable canard can be stalled, primarily by performing what is known as an accelerated stall. Here, the aircraft is put into a very abrupt pull-up that, due to its curved flight path adds an additional 'g'-load to the airplane, increasing its stall speed and forcing the upper surface canard airflow to separate.

One final word on stalling. Always remember that any ultralight will stall at any time provided its angle of attack is too high — no matter what the airspeed, throttle setting, attitude, configuration or 'g'-load!

NOTE: As a beginning ultralight pilot, or even as a seasoned pro, it might be a good idea to rig up an angle or attack indicator of your own. It's simple. Attach a stiff wire or dowel to the aircraft so that it protrudes into undisturbed airflow. Bend the end down 90° and attach a length of string (6-12" long) to it. And there you have it. No matter what airspeed or attitude you're in, that simple string will tell you about your angle of attack!

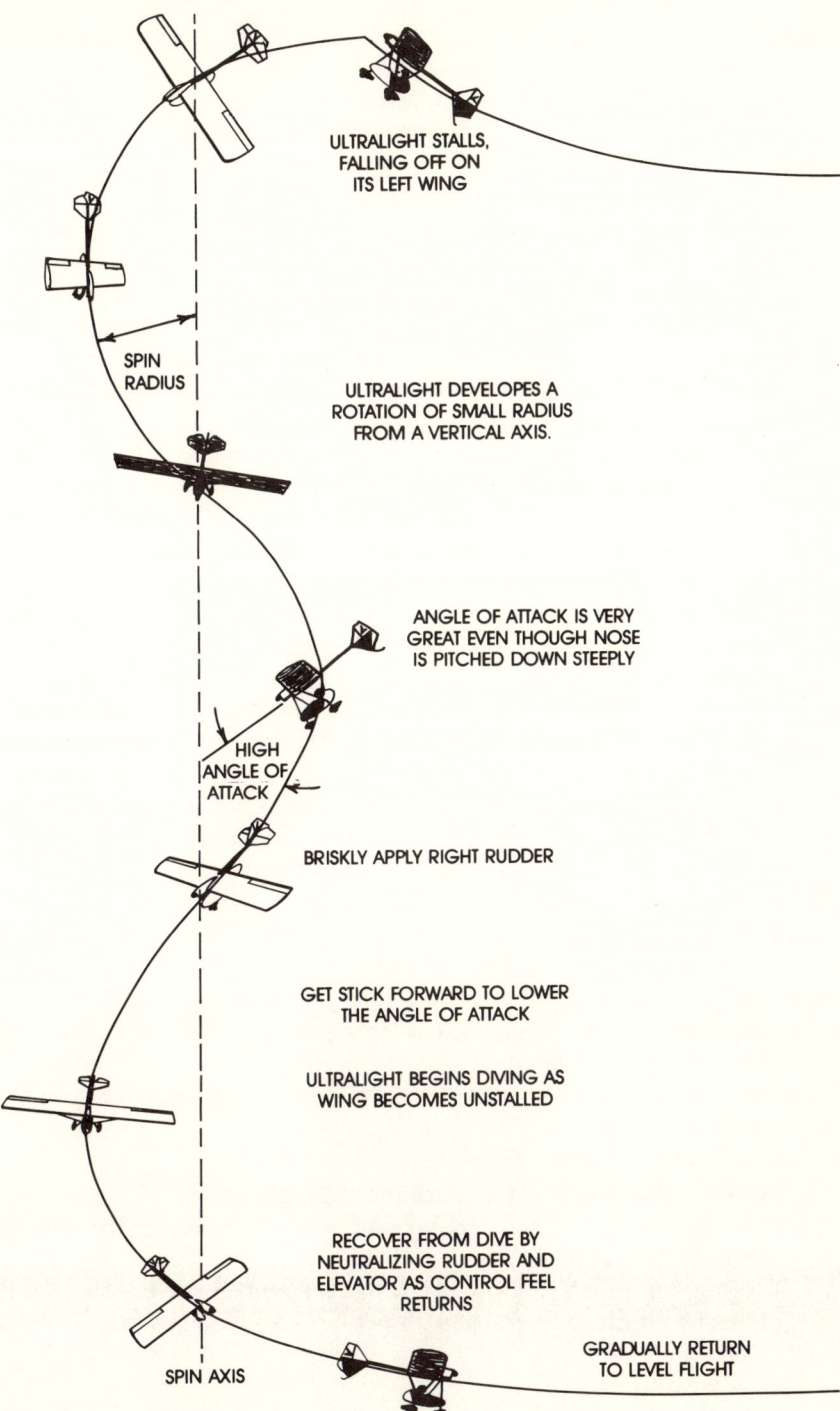

Fig. 7-3. The sequence of a spin. The air vehicle is stalled and rotating a small radius from a vertical axis.

Preflight Inspection

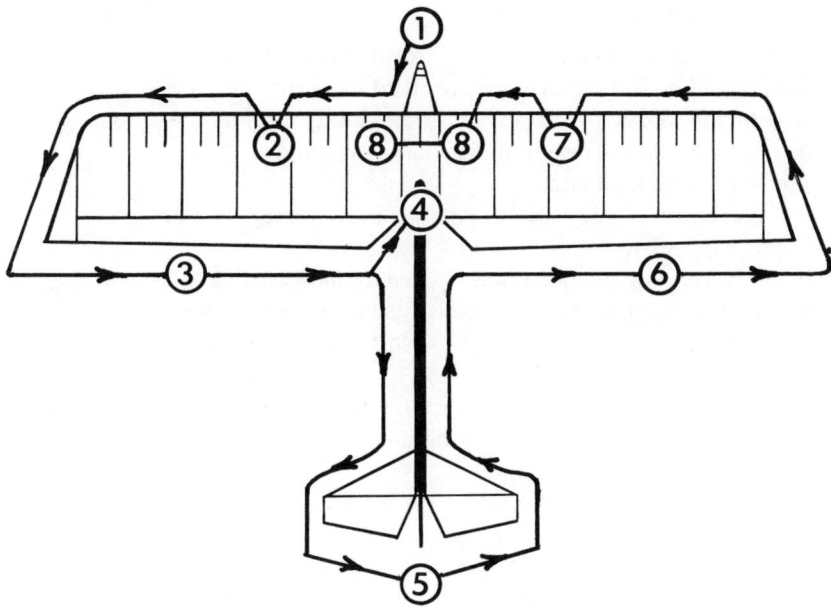

Fig. 2-1. The pre-flight inspection involves a systematic look at everything on the air vehicle.

    vice versa. Right rudder pedal moved forward pulls rudder to right. Left rudder pedal moved forward pulls rudder to left. All control system components secure and operating freely. All pulleys sound and cable safeties in order.
    4. Instruments — Properly mounted. Pitot tube clear and in proper location.
2) Left wing strut
    1. Strut to fuselage anchor point — Secure and safetied.
    2. Strut to wing anchor point — Secure and safetied
    3. Strut — Free of kinks, bends, dents and scratches
3) Left wing trailing edge
    1. Aileron — Hinges secure. Aileron moves freely.
4) Propulsion System
    1. Engine mount — Weldments sound, no cracks. Nuts safetied.
    2. Spark plug — Tight, and wire firmly in place.
    3. Exhaust pipe and muffler — Sound and secure. No cracks.
    4. Drive belt — Properly tensioned with no cracks. Check for wear.
    5. Propeller — Blades free of nicks and cracks. Hub nuts properly safetied.
    6. Fuel — Tank filled with proper ratio fuel/two-cycle oil mixture. Free of water and other contaminents. Fuel lines sound and securely fastened.

Spins

A spin is defined as the auto-rotation of a stalled aircraft around a vertical axis, coupled with a large altitude loss with each revolution. It happens when one wing stalls before the other, as might occur in a turn made too near the stall.

If, for instance, you're in a left turn near a stall, and you decide to level the wings before you lower the nose. The left wing's down going aileron will slow the wing and stall it — rather than raising it as intended. The left wing will drop further still and set-up a rotation to the left, as the nose drops below the horizon — a spin will be started.

Now, even though the nose is below the horizon, the wing is at high angle of attack — due to the vertical component of velocity — and it is as stalled as it can be. At this point the ailerons are quite useless and will do nothing to correct the situation. As in all other phases of flight, to be flying requires that the wing be at an unstalled angle of attack, and that means — get the stick forward. Yes, even though you are pointed down already, the stick must come forward in order to lower the angle of attack. If the engine is at an advanced rpm, the throttle must be closed immediately.

Extreme care must be taken in recovering from the dive that results from a spin recovery. After the spin rotation stops, back pressure is to be applied *slowly* in order to restore level flight. If the stick is pulled back too abruptly, the ultralight's airframe could be overstressed — tearing the wings off! Excessive back stick could also cause the aircraft to enter a secondary stall or worse yet, another spin.

A spin could also be stopped by applying opposite rudder, but it depends on the design. It is not known how thoroughly some ultralights have been spin tested and you'd do well to inquire at the factory or dealership to find out. If the particular design you're interested in has been spin tested, the manufacturer will have a specified method of recovery. If it hasn't been spin tested, don't volunteer to be the test pilot — fly to avoid spin potential by maintaining proper airspeed, or more correctly, proper angle of attack at all times. Some designs may be claimed as "spin proof", but why tempt fate by flying too near the stall speed.

Putting it all together then, the standard spin recovery technique can be stated as follows:

1. Briskly apply rudder against direction of rotation.
2. Push the stick forward to reduce the angle of attack.
3. As rotation stops and control pressure is regained, center the rudder and level the wings.
4. Recover from the resulting dive with very gentle back pressure, and gradually return to level flight.

8

Landing and the Traffic Pattern

Landing is often said to be the most difficult part of flying, followed by takeoff, then actual cruising flight.

The actual steps involved in making a landing in a standard left hand traffic pattern, are as follows:
1. Choose the runway.
2. Enter the pattern 45° with respect to the downwind leg, at 300 feet AGL.
3. Turn 90° at end of downwind leg and fly base leg as you descend, with reduced power.
4. Turn 90° at end of base leg and set up final approach glide.
5. Determine your perspective and select your touchdown point, and glide toward it.
6. Break the glide as you approach the ground.
7. Level off 2-3 feet above ground.
8. Hold aircraft off as it loses flying speed.
9. Bring the nose up to landing attitude.

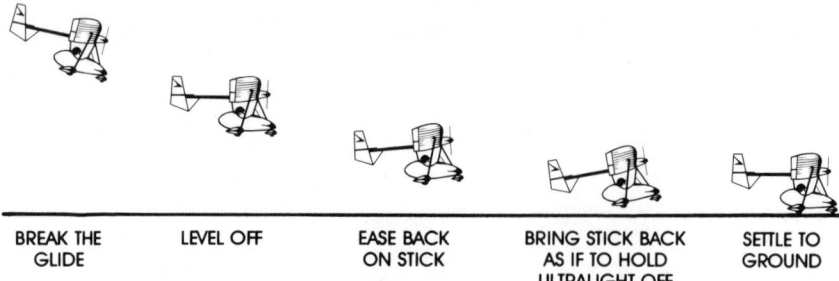

BREAK THE GLIDE — LEVEL OFF — EASE BACK ON STICK — BRING STICK BACK AS IF TO HOLD ULTRALIGHT OFF — SETTLE TO GROUND

Fig. 8-1. The sequence of a normal "full stall" landing.

10. Sink to the ground for a touchdown.
11. If a taildragger, hold the stick back so the tailwheel is effective. If a trigear, hold the stick forward to plant the nosewheel firmly on the ground.

NOTE: Steps 9 through 11 could be modified to allow touchdown above stall speed. This would be a "wheel landing" where the aircraft is "driven-onto-the-ground". The reason being to stay well above stall and not be susceptible to stalling out from an unanticipated gust.

Now let's analyze each step of the landing procedure.

Choose the Runway

Before choosing the runway, it is necessary that you determine the wind direction. If there's traffic at the field, their takeoff direction will tell you which direction to land. If you're out in the middle of nowhere, you could look for telltale signs, such as drifting smoke or flags. If there's a windsock, check it out. You'll be landing into the free end.

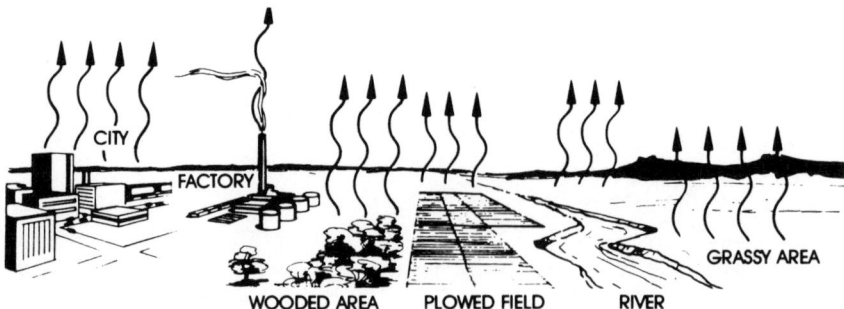

Fig. 8-2. The earth's surface heats unevenly, giving rise to convective currents of varying strength.

Enter the Pattern

A traffic pattern is a rectangular course flown over the ground at certain altitudes in a systematic procedure, allowing consistent landings to be made. The pattern for ultralights looks like that shown, as recommended by the AOPA.

Landing and the Traffic Pattern

In conventional aviation, it is common practice to enter the pattern at 45° with respect to the downwind leg, and it makes good sense to do this with an ultralight, too. This entry might occur at 300 feet above ground and 1000 feet parallel to the runway, at an airspeed of 45 mph. Fly the downwind leg maybe 250 feet downwind from the end of the runway and turn 90° onto the base leg. Come back on the throttle to say 1½ times idle speed and begin the base leg at 40 mph.

The Final Glide

Turn 90° to final at maybe 100 feet above ground and slow to 35 mph. Line up with the runway and use throttle as required. Remember, the engine could quit at any time so it's better to have a little excess altitude than not enough, in case you have come in "dead stick" (propeller stopped!)

It is important to remember that on final you are actually a glider and, unlike cruising flight, your nose will have to be lower. The nose will be aimed below the horizon to maintain that 35 mph glide. If this isn't the case and you're at a level attitude, the airspeed will drop below 30 mph and the airplane could stall, which would be disastrous.

It's a real good idea to keep an eye on the airspeed, as ultralights tend to lose it rapidly with a pitch up. Also, remember that you're slower now than when cruising, and the controls won't be as responsive. More anticipation is necessary. Come in a little faster if it makes you feel more comfortable. It's better to have too much speed on final than too little. Look at it as an extra margin against a gust stalling you out.

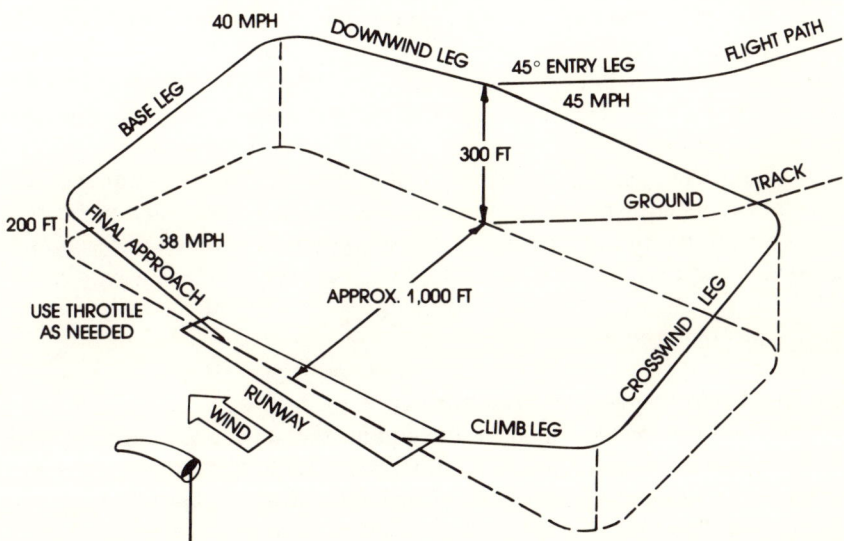

Fig. 8-3. An ultralight traffic pattern as practiced by many pilots.

Since gliding is different than cruising, you should climb to an altitude of say 1,000 feet, and practice gliding with the engine slightly above idle. And remember to "clear" the engine (i.e., give it a shot of ¾ throttle) every 30 seconds to keep it warm. If it gets too cool, it could quit — then you would be flying a glider for sure, and be forced to land. Pick a point on the ground, and practice gliding toward it. Improve your skill with the safety of altitude.

Remember, if the point you aim for is neither rising nor falling in your field of view, you'll reach it. If your aiming point is rising, you'll undershoot and not reach it. You'll have to add power to maintain your present altitude or to rise. If your aiming point is dropping, you'll overshoot and touchdown beyond it. This will require going around the pattern again or slip it in, as explained in the next chapter.

Level Off

After you have mastered the glide and can accurately control it to reach the desired point, you're ready to brake the glide and level off.

In some respects, leveling off is much like parallel parking a car. If you enter the space at too much of an angle, you'll hit the curb. And so it is with leveling off. And, as in other maneuvers, a certain amount of judgment and anticipation is required. You must be able to assess your sink rate and anticipate it being zero when you touch down.

When you are about 5-10 feet above the runway, you should break the glide. This requires coming back on the stick, while maintaining a good margin above stall speed. This will slow your rate of descent and round-out the glidepath. When you're about 1-2 feet above the ground, the round-out should be completed and you should be flying parallel to the runway.

If you level off too high, the ultralight could stall before it touches down, dropping and damaging the airframe and injuring the pilot. If you level off too low, the aircraft will strike the ground too hard and bounce back into the air, only to stall out on the way up. Full throttle and a go-around might be the only solution.

A good way to judge the leveling-off altitude while maintaining directional control at touchdown, is to maintain a visual focus beyond the touchdown point toward the end of the runway. Focusing too close will blur objects and cause delayed reaction or over controlling. Focusing too far away, beyond the runway, will cause altitude judgement to be off and the nose may be flown into the ground.

Losing Flying Speed to Touchdown

After you have leveled off and are skimming the runway, ground effect will be felt and tend to cause you to float. Be ready for it. Nevertheless, you will eventually sink and your speed will drop to very near stall speed, while your height will diminish.

As you come closer and closer to the ground, keep coming back on the stick, as if you were trying to hold the airplane off. Come back gradually on the stick, or you may zoom back up into the air, only to stall out. This zoom is called ballooning and is certainly to be avoided. At any rate, keep coming back on the stick as you sink and a good landing should result. This technique describes a so-called full-stall landing, as you are at or near stall when you touchdown. It is fine for calm and very light wind landings.

Wheel Landing

When the wind is above about 5 mph and/or gusting and/or not straight down the runway, a full landing could be dangerous. In these situations, it is often wise to carry extra airspeed (as much as 15-20 mph above stall) as an insurance policy against having a gust cause you to lose flying speed and stall.

In a wheel landing, you effectively level off at the instant you touch ground. A taildragger would touch on its main wheels only and keep the tail up. A trike gear could touch all three wheels at the same time. What you are doing is essentially driving the ultralight onto the ground at above flying speed. The stick must be held forward as soon as you touch, so as to prevent ballooning.

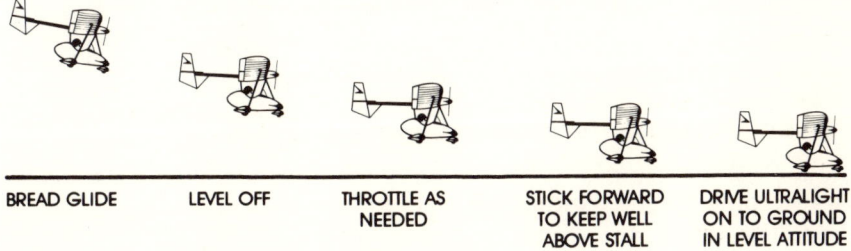

BREAD GLIDE | LEVEL OFF | THROTTLE AS NEEDED | STICK FORWARD TO KEEP WELL ABOVE STALL | DRIVE ULTRALIGHT ON TO GROUND IN LEVEL ATTITUDE

Fig. 8-4. The sequence of a "wheel" landing.

Judgment and Concentration

Landing is a matter of perspective. How do objects appear from your point of view? What angles do things appear to present? A good example of how perspective can be used in landing is for you to look at an object while standing and then squat down. The object will appear to change its shape somewhat. The angular relationships will change as you move the level of your eyes for even a couple of inches! Perspective is a powerful tool for judging a landing.

It shouldn't be necessary for you to look at the wheels of your ultralight to make a good landing, but you may want to at first. Look all around, especially to the front and focus on your aiming point. Notice if it is rising or falling or steady. After a while, you will be able to make perfect touchdowns without watching the ground directly beneath your wheels.

Landing is also a matter of concentration. Concentrate on what you are doing and don't allow yourself to be distracted from your task. Concentrated effort can go a long way in learning and save lots of time besides. (Caution:

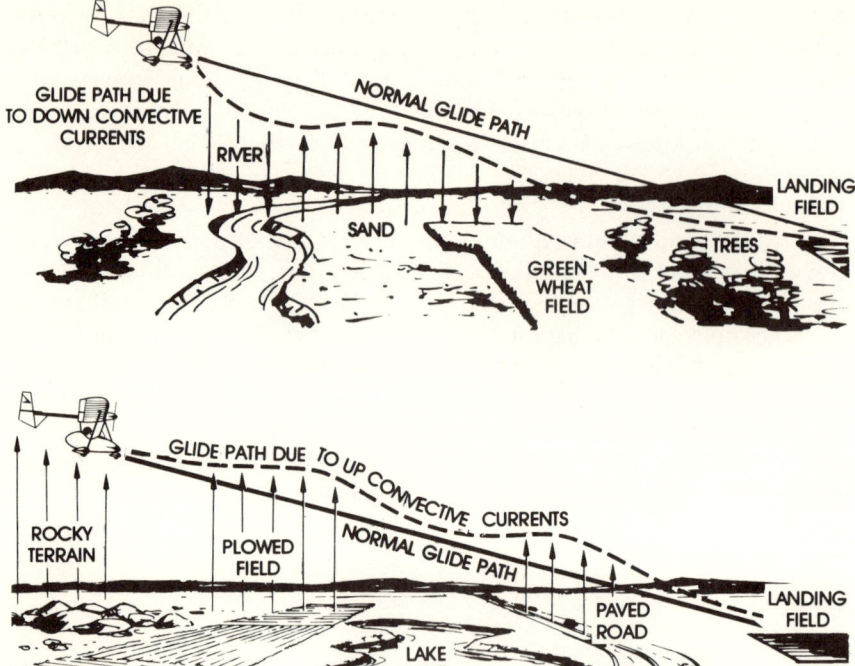

Fig. 8-5. Convective currents have a definite affect on the glide path of an ultralight.

while you are concentrating, don't forget about other traffic. A mid-air collision can ruin your whole day.)

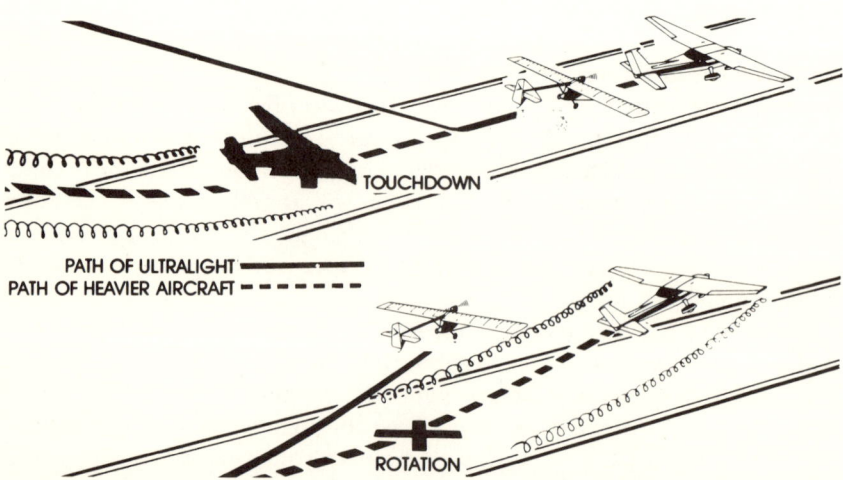

Fig. 8-6. When operating from a runway with a heavier aircraft, always land in front of its touchdown point, and takeoff before its rotation point. Better yet, don't operate with heavier aircraft at all. There wingtip vortices could easily roll an ultralight over on it's back!

Landing and the Traffic Pattern

LANDING IS A MATTER OF PERSPECTIVE, not depth perception. Perspective is different for different ultralight designs, and winds. Greater headwinds and "dirtier" ultralights have the landing field appearing more like No. 1. No. 2 is about right for many ultralights under normal circumstances. Calm conditions and "cleaner" designs may favor No. 3.

1. TOO LOW TO REACH WITHOUT POWER.
2. POSITIONED FOR STRAIGHT-IN GLIDE.
3. TOO HIGH. GO AROUND.

HERE'S HOW TO JUDGE IF YOU'LL LAND ON THE DESIRED SPOT. If your intended landing spot appears to move up, you won't reach it. If it appears to move down, you'll overshoot. If it doesn't move at all, but only grows in size, you're on the right glide path, and you'll touchdown on the intended spot.

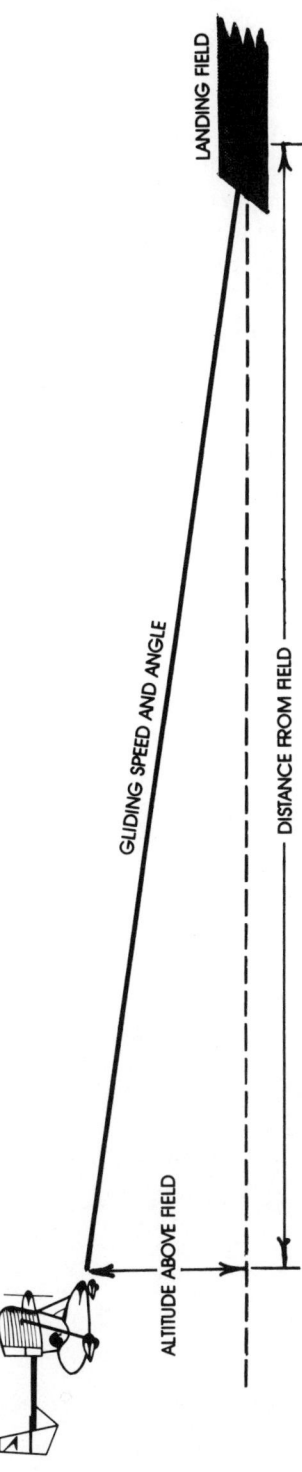

Fig. 8-7. Factors that determine the correct glide path to make a landing in the desired spot.

Post Touchdown Taxi

Once you have touched down you must continue to fly the aircraft, especially if you did a wheel landing. You'll be in a very high speed taxi situation requiring all of your attention. Until you're below stall speed, the airplane is still flying, and the flight controls will remain functional. "Fly" the ultralight on the ground until the speed drops below stall, then switch over to your taxi techniques. If you have to turn, remember to do it gently, or you may find yourself in a ground loop.

Soft Field Landing

A soggy field can be a real hazard to a landing ultralight. If the wheels would dig in during touchdown, you'd be in for a nose-over and its consequent damage and injury. The basic technique requires a flat approach, slowed as much as possible. The touchdown should be at or below your normal power off, free-air stall speed. After you do touch, maintain an aft stick to slow weight transfer to the field and minimize the chance for a nose-over.

9

Sideslipping and Other Ways to Lose Altitude

Suppose you are too high on final approach and your engine has just stopped and there's no way you have enough altitude to go around. You simply must set it down, since the field below you is the only one within gliding distance — in the middle of a forest of tall trees. What do you do? Dive it in? No. You'd have too much speed near the ground and float in ground effect to the end of the field and crash into the trees at the other end.

The solution to the above scenerio is called the sideslip — a deliberate cross-controlling of your ultralight. It will allow you to decrease your glide ratio to perhaps half of what it normally is, while still maintaining good airspeed. You'll be able to drop 2 feet in 10 instead of one foot in ten, for example.

Having set up your approach to the field in a normal glide, you are ready to slip. Depress the right rudder pedal gradually and keep it depressed, while

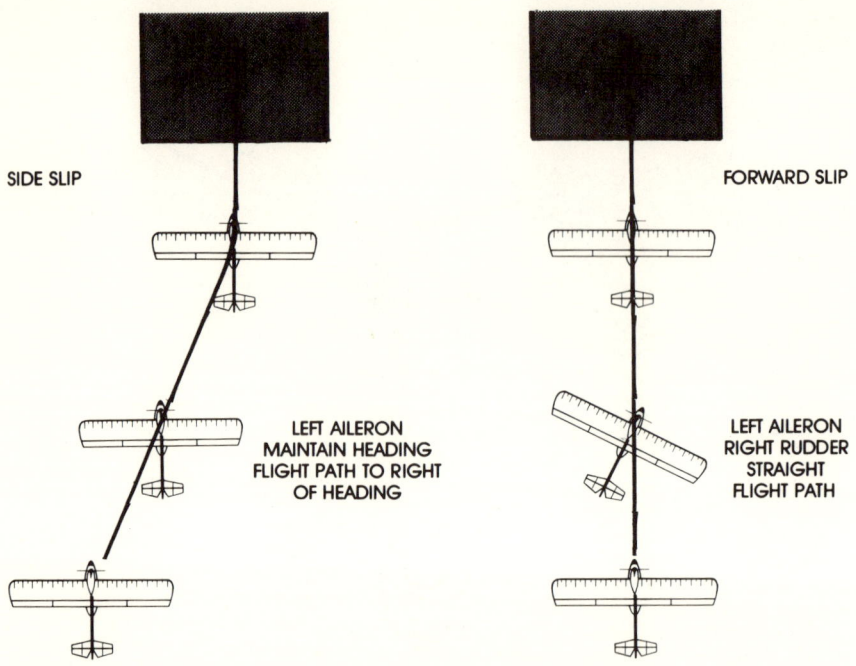

Fig. 9-1. Side slip and forward slip exposes the air vehicle's side to the flight path, increasing drag for a steeper descent, when necessary to lose altitude in a short distance.

at the same time feeding in some left stick and holding it. The ultralight will assume a right heading with the left wing down, while the glidepath will steepen at the same glide speed.

By increasing rudder and aileron deflections as above, the sideslip can be increased, steepening the glide still more, with a little increase in glide speed. The nose will be pointed further right, while the left wing will be lower still.

To control the direction of the sideslip, think about how an aircraft is turned, as discussed earlier. If you'd like to direct your course to the right, to correct for drift for instance, simply raise the left wing a little, thereby directing some lift to the right. If you wish to turn left, let the stick where it is, and release some of the right rudder you've been holding.

When you get down to the point where you're ready to break the glide, release the rudder and bring the stick back to neutral. This will straighten the aircraft so that it is headed and flying in the direction of the original glide. Now make corrections as necessary to straighten the glidepath with respect to the field and touchdown going straight ahead. This is important, for if you have sideways speed, the landing gear could break on touchdown and/or the airplane could cartwheel onto its side. Always touchdown with forward speed only.

NOTE: The foregoing described a left sideslip. It is also possible to do a right sideslip by pressing left rudder and holding right aileron.

Sideslipping and Other Ways to Lose Altitude 81

Why a Sideslip Works

At this point you might be puzzled as to why a sideslip can work. After all, aren't the ailerons supposed to roll the airplane? A sideslip works because the rolling force of the ailerons is counterbalanced by the extra air pressure under the low wing which has a sideways velocity component tending to pick it up. And, this same sideways velocity also strikes the fin, counterbalancing the yaw produced by the rudder. So, a state of equilibrium is reached with the airplane flying in a slip. And, since the controls are deflected so abnormally, and areas of the airplane are exposed to abnormal airflows, the total drag rises, steepening the glidepath.

How to Slip a Turn

If the occasion arises that you must lose altitude more quickly than normal while turning, it is useful to superimpose the sideslip technique on to your turn. Start by entering the turn as normal, using the appropriate amounts of pitch and yaw with the bank, so it is coordinated. Once that's established increase your bank slightly while applying some opposite rudder (i.e., diminish the amount of rudder into the turn). That's a slipping turn. To get out of it, simply restore the rudder and stick so that the turn is once again coordinated.

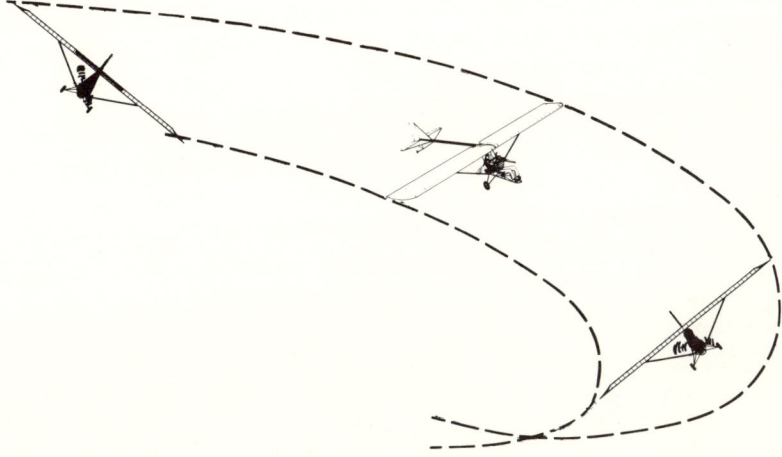

Fig. 9-2. A slipping turn has the rudder held out of the turn.

Fishtailing

Another way to steepen the glidepath is by fishtailing — the application of rudder, alternating right and left. Done without any aileron inputs, this results in a series of alternating slips, while maintaining the desired direction of flight. It's not as effective as a fully developed slip, but it is another method of helping you get down as desired.

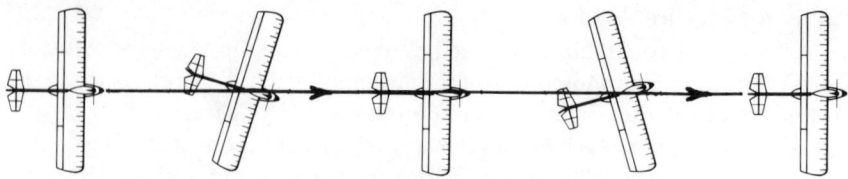

Fig. 9-3. Fishtailing is a series of linked left and right rudder applications, done without aileron. This results in a succession of alternating forward slips.

Tip Draggers

Some ultralights are also equipped with tip draggers or tip rudders. These not only yaw and induce banks when deflected independently, but they can also be used to control the glidepath when they are deflected simultaneously. The basic technique is to apply some deflection to both draggers while on final. That way, if it looks like you're going to undershoot, let them both go, and your glide will increase. If you're too high, deflect them both as needed to steepen the glide.

10
Crosswind Operations

Whenever you choose to fly when there is wind, you will encounter a crosswind at some point in the flight, be it in cruising, turning, takeoff or landing. Or, you might have an engine failure and be forced to land in a field that is in a crosswind. Now, since most ultralight pilots will not restrict their flying to calm conditions, it would be wise to learn how to handle crosswinds.

The Technique of Crabbing

Suppose you want to fly from your field over to a friend's farm, which is due south, and there's no wind blowing from the west. If you were to fly south, you would never reach the farm — the wind would have blown you east as you traveled south! This is called drift and you must crab to overcome it. For example, if your cruising speed is 40 mph, the wind is at 5 mph and your friend's farm is 40 miles away, in an hour's time you would end up 5 miles east of his farm!

In order to reach your friend's farm, your heading must actually be west of south, i.e., slightly into the wind. The greater the crosswind, the more west of south your heading must be. The ultralight would be flying sideways with

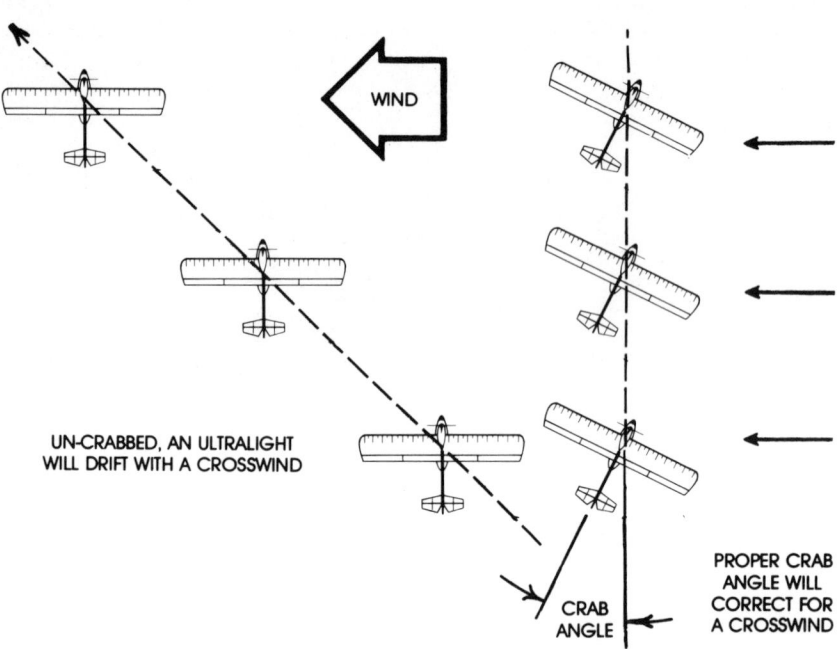

Fig. 10-1. You must crab to correct for wind drift.

respect to the ground, and you'd be aiming at a point on the ground, left of the airplane's centerline.

Being in this situation for the first time, you immediately surmise that the solution to this problem is to kick in some right rudder. So you try it — but it doesn't work! Sure, the aircraft yaws to the right, but it still drifts with the wind. Even if right rudder is held continually, the airplane would only make circles, and skidding ones at that, that would still drift with the wind.

Thinking again, you recall reading somewhere in this book that ailerons roll the airplane and tilt the wing's lift in the direction of the stick. Perhaps you can tilt the lift into that pesty crosswind and make the airplane go where you want it to? You try, and viola, it works! In effect, you are turning to compensate for drift, and that's the way it's done. A crab is established simply by a turn, done in the normal way with aileron, while the rudder is used to correct any adverse yaw. Once you are tracking over the ground as desired, the controls are neutralized, and the aircraft will continue to crab. If the crosswind changes in direction or intensity you will, or course, need to make appropriate heading corrections in order to maintain your desired track.

Crosswind Takeoff

While an ultralight is in contact with the ground it, of course, must be headed in the direction it is going. Sideways motion could damage the landing gear or tip the aircraft over and into a ground loop.

Crosswind Operations

The basic technique used in a crosswind takeoff is to hold stick into the crosswind to tilt the lift toward it, while applying appropriate opposite rudder to counteract any weathercocking tendency. Sound familiar? It should, because that's exactly what a slip is. Once airborne, you can control the drift by maintaining proper aileron and necessary rudder to correct for any unwanted yaw, and that is a crab.

Crosswind Landing

Now that the cat is out of the bag, landing in a crosswind should be easy. On your final approach you can either crab or slip, depending on whether you need to steepen the glide to get down. Then, at the end of the runway, before touchdown, straighten out by going into a slip and you'll touch down headed in the direction you're going — straight down the runway.

NOTE: One nice thing about ultralights in regards to crosswind takeoffs, and landings is that exposure to the event is very limited. Whereas a typical light plane might require 800 feet of ground roll, an ultralight needs only around 100 feet. While exposure to crosswind conditions is minimal, proper technique must be applied to operate successfully.

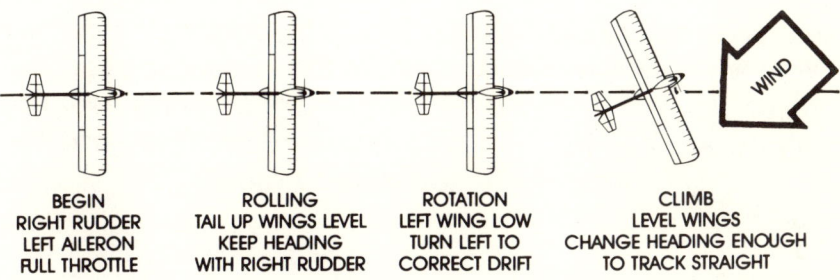

BEGIN	ROLLING	ROTATION	CLIMB
RIGHT RUDDER	TAIL UP WINGS LEVEL	LEFT WING LOW	LEVEL WINGS
LEFT AILERON	KEEP HEADING	TURN LEFT TO	CHANGE HEADING ENOUGH
FULL THROTTLE	WITH RIGHT RUDDER	CORRECT DRIFT	TO TRACK STRAIGHT

Fig. 10-2. The technique for a crosswind takeoff.

"A crab is established simply by a turn, done in the normal way with aileron (spoiler), while the rudder is used to correct for adverse yaw. Once you are tracking over the ground as desired, the controls are neutralized, and the aircraft will continue to crab."

11
Instrumentation for Ultralights

While it is entirely possible to fly an ultralight solely by-the-seat-of-your-pants and the wind in your face, it is strongly recommended that you employ at least minimal instrumentation. Due to various accelerations not encountered in surface travel, and just plain unfamiliarity with being in the air, your senses can sometimes lead you to believe something you should not. For these reasons, and also for accuracy and safety in flight, there are four primary instruments that you should have on board: airspeed indicator, altimeter, tachometer and compass.

Airspeed Indicator (ASI)
The airspeed indicator tells us how fast we are moving relative to the air, and it's the most important instrument of the group. Since the stall speeds of ultralights are so low, it can be difficult to judge exactly when the stall will occur, by such a method as sound. In this regard, an airspeed indicator is an essential piece of equipment. At best though, it is second to angle of attack as a stall indicator — but it's the best device available to relay the information.

Fig. 11-1. The Orbit instrument package is a streamlined unit containing the three primary flight instruments. Contact: Orbit Aircraft Instruments, Ltd., 110-904 St. James Street, Winnipeg, Manitoba R3G 3J7.

An airspeed indicator is very useful in adjusting throttle and stick to make the airplane fly at a prescribed speed, be it for efficiency, safety or performance. As discussed elsewhere, there are *fourteen distinct speeds* at which you can fly, and only the ASI will tell you when you're there.

The ASI is also extremely useful when flying near the ground. If, for example, you are forced to land downwind for some reason, your

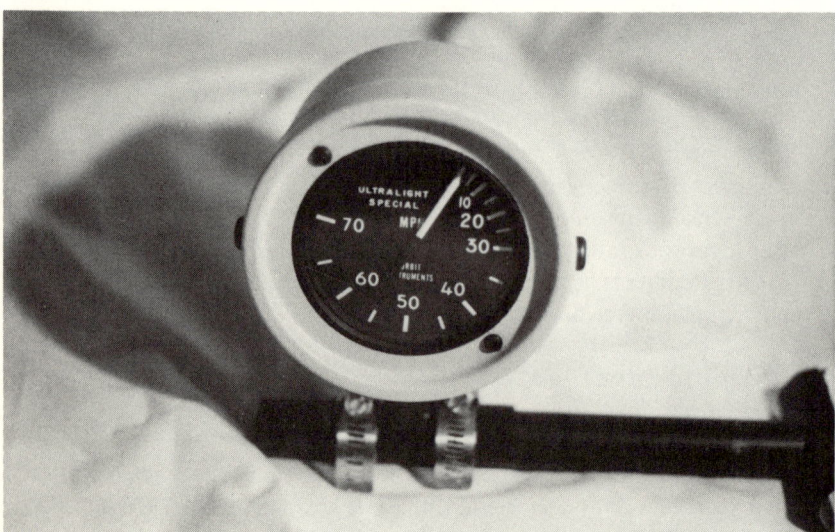

Fig. 11-2. The Orbit airspeed indicator is specially designed for the needs of the ultralight enthusiast.

groundspeed would be your airspeed plus the wind speed. In this instance, the ground would be rushing by faster than normal and unless you watched the ASI, you'd probably slow down and stall. Sure, your groundspeed was high, but it doesn't count. It's airspeed that enables you to fly.

How the ASI Works

The standard aircraft ASI operates on the difference in pressure between the impact pressure caused by forward motion and the static pressure of the atmosphere. These pressures are picked up by a pitot-static tube located in front of a wing or strut, where it feels the freestream airflow. These pressures are then transmitted via two tubes to the ASI itself, where the pressure difference is translated into mph.

An ASI used on some ultralights consists of a vertical tube with a foam ball inside that rises with increasing airspeed. The inside of the tube is cone shaped, allowing the ball to rise higher with increasing impact pressure. Graduations on the tube indicate mph. Other ASI's would be a small propeller attached to a small DC motor connected to a millovoltmeter calibrated in mph, or a simple spring loaded vane that angles back with increasing speed. Helicopter ASI's are also being used on ultralights, because of their sensitivity to the low speed range we fly in.

Fig. 11-3. A helicopter airspeed indicator can be used for ultralights, but they're expensive.

Another popular ultralight ASI is a unit, manufactured by Airguide, that also includes a compass. The handle is removed and the instrument is mounted to the front of the aircraft, where it sees unobstructed airflow. It is extremely important that the air inlets be unshielded and the unit must point into the relative wind. If the airflow into the device is altered, by say a cowling, the airspeed readouts would be incorrect and could spell disaster. It must be calibrated, as it is mounted on the aircraft, with a known source.

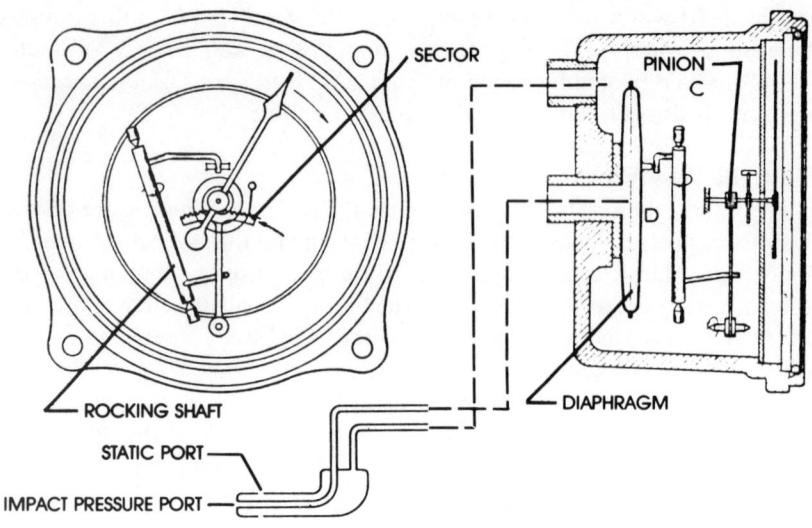

Fig. 11-4. The inner workings of an airspeed indicator.

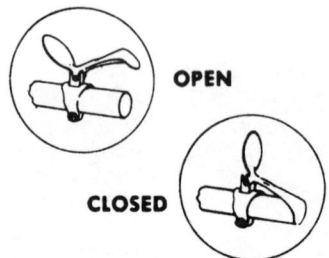

Fig. 11-5. A pitot tube inlet protector makes for a wise addition to an ultralight using a pitot static airspeed indicator system. It opens with forward motion and stays closed when not, keeping out insects that could plug the system.

Altitude Correction

As you gain altitude, the air becomes less dense, which means for a given true airspeed, the impact pressure is less. To compensate for this, we must add to the indicated air speed about 2% for every 1,000 feet of altitude. At 5,000 feet you would need to add 5 x 2% = 10% to your indicated. If you're indicating 50 mph, your true airspeed would be 50 + = 55 mph.

Now you ask, does stall speed change with altitude? And, the answer is yes and no. Fortunately, the *indicated stall speed* doesn't change with altitude, whereas the *true airspeed* does. But, this won't matter to you — watch the ASI. As we said recently, the air is less dense at altitude therefore, to get the same impact pressure — which is really what matters to the wing — you must fly 2% faster *true airspeed* for every 1,000 feet of altitude. The ASI will still *indicate* the same value for stall speed, no matter what the altitude.

Instrumentation for Ultralights

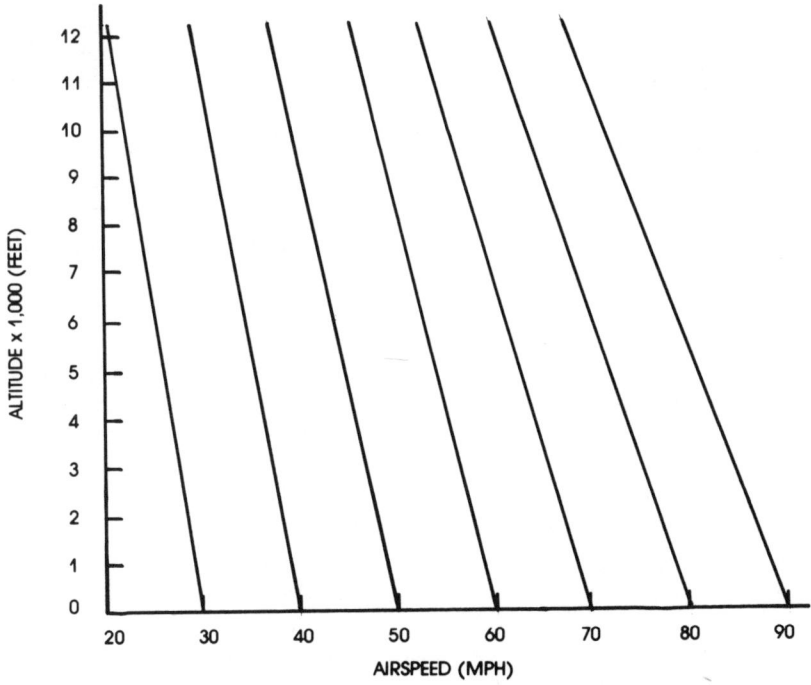

Fig. 11-6. This graph enables you to correct indicated airspeed for the affects of altitude.

Fig. 11-7. How the instruments and controls look on a two-seater ultralight. Flight instrument cluster is on right, while engine is monitored on left panel. This trainer is equipped with full dual controls.

The Altimeter

Unless you are quite close to the ground, it is nearly impossible to accurately judge altitude. For this and reasons of safety, we have the altimeter.

An altimeter has a clock-like face and hands, as well as a small window scale that can be set to read local barometric pressure, with a small knob located on the lower left corner of the case. The hands rotate clockwise for altitude gains and vice versa. The large hand indicates altitude in thousands of feet, while the small hand points to hundreds.

How an Altimeter Works

The altimeter is actually an aneroid barometer that is calibrated to read in feet. It makes use of the fact that air density decreases with altitude. A wafer-like capsule, the aneroid, is a thin wall piece sealed under a vacuum. A small spring keeps it from collapsing under outside air pressure. The meter case itself is under normal outside static air pressure. When altitude is increased, the outside air density decreases, and the aneroid-spring arrangement expands. Through a system of levers and a chain, the small expansion is magnified and turned into a rotary motion of the pointers. Thus is altitude indicated.

For local work, some pilots prefer to set the altimeter to zero, no matter what the barometric pressure, for zero will be the level of the field and serves as a convenient reference.

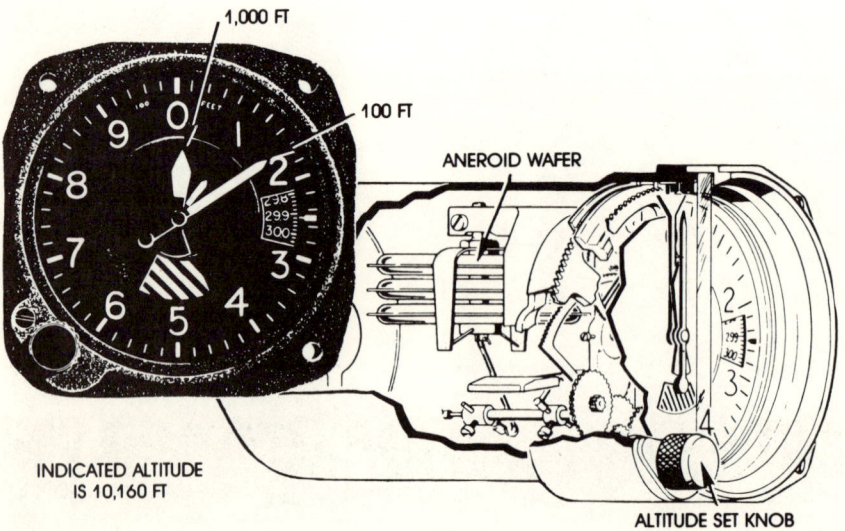

Fig. 11-8. A typical altimeter and its inner mechanism. While altimeters can be had with one, an altitude set knob is highly recommended. USAF.

Instrumentation for Ultralights 93

Fig. 11-9. A rate of climb indicator is not necessary for normal flying, but can be a nice addition to any instrument panel. An instantaneous rate of climb, or variometer, is essential for soaring flight.

The Compass

Unless you are planning to fly only around your local area, a compass is a wise investment. Besides, it is nice to have anyway, as it can be used to improve your flying skills.

An aircraft compass is very similar to a standard issue hand held type, with a few modifications. The main difference is that the numbers are on a pivoted card immersed in a damping fluid. A vertical line (lubber line) across the face of the compass, crosses the direction you are headed. The card points north at all times and the lubber line, along with the airplane, rotates around it. The "swing-of-the-card" may be confusing at first, but if you remember that the lubber line, along with your aircraft, rotates around it, the confusion should vanish.

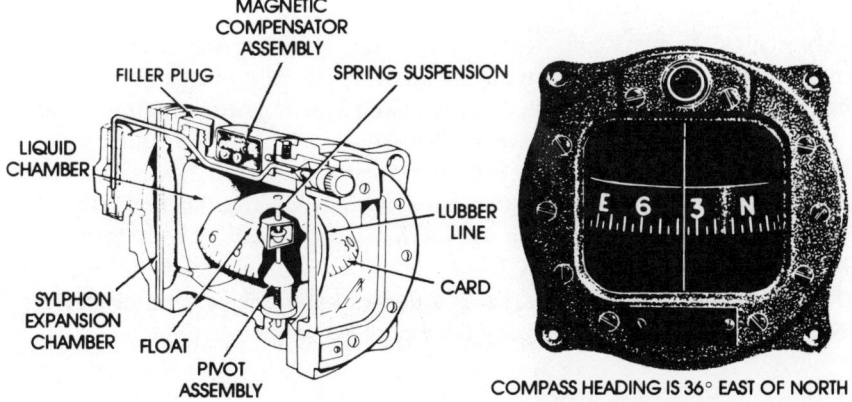

Fig. 11-10. The aircraft compass and its inner workings. A compass is a must for cross-country flying. USAF.

Standard aircraft compasses have small compensating magnets to adjust the card to magnetic north under the influence of any iron and steel that may be on the airplane. Most ultralights, however, don't have this problem, since aluminum construction prevails. However, the engine is a large mass and it contains steel, which may affect the card.

There are two in-flight situations that affect the reading of a compass: northerly turning error and acceleration error.

Northerly turning error is the most pronounced. Due to the vertical component of the earth's magnetic field, when a turn is made to either side from north, the compass lags or even initially indicates a turn in the opposite direction. This is due to the north seeking end of the card dipping to the low side of the turn. Because of this, it is necessary to begin your roll-out about 10° before east or west or at 80° or 280°.

If you are heading south and turn east or west, the effect is reversed, i.e., compass leads the turn. In this case, start your roll-out only 5° ahead or at 95° and 265°.

Acceleration error is also caused by the vertical component of the earth's magnet field and the pendulum type mounting of the card — acceleration causes the card to tilt upward, deceleration causes it to tilt downward — both of which are most noticeable heading east or west. If you accelerate while on either heading, the compass card turns northward whereas if you decelerate, it'll turn southward.

What this all means is that the compass gives an honest magnetic reading only in straight and level flight. A word to the wise is sufficient.

When using a compass for navigational purposes you must be aware of the fact that the magnetic north pole does not coincide with the geographic north pole, and that the earth is not uniformly magnetized. Because of this, it is necessary to add or subtract several degrees from your compass indication, depending on your location. The proper amounts are shown on sectionals (charts used in navigation) as Isogonic lines, or lines of equal magnetic variation.

In central Pennsylvania, for instance, the Isogonic line reads "10°W", meaning the compass will actually be pointing 10° west of true north. To fly due east, you'd have to add the 10° to 90° and fly a magnetic heading of 100°!

There is one line called the Agonic line, for which true north and magnetic north coincide. It runs roughly from the Great Lakes south, through just east of Florida. In this region, the compass will indicate true north as 0°

On the western side of the Agonic line, it is necessary to subtract the Isogonic line variation values from what your compass reads, in order to determine your true heading. To fly due west, you'd have to subtract from 270°, depending on your locale's Isogonic line variation value.

The Skid/Slip Indicator

A skid/slip indicator could also be useful as an aid in determining the coordination of a turn. It's simply a metal ball in a curved glass tube

Instrumentation for Ultralights

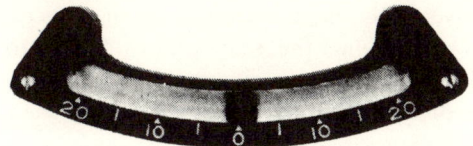

Fig. 11-11. The skid/slip indicator tells you if your turn is coordinated (ball centered) or not. The ball goes outside the turn in a skid, and inside the turn in a slip. The technique in coordinating a turn is to "step-on-the-ball", with appropriate rudder pedal.

containing a damping fluid. As long as you are flying straight and level, or in a coordinated turn, the ball will be centered. The indicator does not show the bank angle. It only shows you if your turn is coordinated. When the ball goes to the outside of a turn, you are skidding. When the ball falls to the inside of a turn, you are slipping.

The ball can be centered by appropriate application of rudder. The thing to remember is to "step-on-the-ball." For instance, if you're skidding to the outside of a left hand turn, the ball will be to the right of center. To coordinate the turn, all you need do is apply right rudder to bring the ball back to center. Alternately, you could also have applied more left stick to increase the bank angle, halting the skid. Be careful not to bank too steeply for the airspeed you're at, or a stall might result.

Turn and Slip Indicator

A further sophistication of the skid/slip indicator is the turn and slip indicator, sometimes referred to as the turn and bank. It was the first gyroscopic instrument developed for aircraft, and works on the principal of precession. Its usefulness in an ultralight is quite optional, but nice to have, especially if you're a licensed pilot used to having one in front of you.

The instrument consists of the skid/slip inclimmeter "ball" and, a needle for turn rate. The ball works on gravity and centrifugal force and is centered when a turn is coordinated. The needle however, is connected to a small electrically driven gyroscope rotating on an axis perpendicular to the aircraft's longitudinal axis. The cage around the gyroscope is pivoted parallel to the aircraft's longitudinal axis. When a turn is made, the force of precession causes the turn needle to move off center, depending on the rate (degrees per second) of turning.

When the needle is positioned under one of the white arrows, the turn is referred to as a "standard rate turn" of 3° per second. In other words, the standard rate turn would allow you to make a 360° turn in 120 seconds, or 2 minutes. With the needle between center and a white arrow, the turn is half the standard rate, resulting in a 360° turn in 4 minutes. The instrument eliminates guess work, giving more precision to your turns.

While a turn and bank indicator is not necessary to safely and proficiently fly an ultralight, it adds a nice touch to any panel, while enhancing your flying skills. The instruments could also prove extremely valuable if, for instance, you inadvertently enter a cloud (which is illegal and extremely dangerous). But, if you do enter that scary grey world of zero visibility, a turn and bank will tell you if you are turning or not. Without it, you would most likely enter a spiral dive, possibly over stressing the air vehicle and breaking it up.

NOTE: The skid/slip indicator alone will not help you in a cloud. It only tells if a turn is coordinated—the ball is centered in a coordinated spiral dive!

Fig. 11-12. The gauges needed to properly monitor an ultralight engine's performance. Those shown are standard electro-mechanical types.

The Tachometer

Your engine speed is a primary piece of information, as your life as well as your engine's depends on it. The tachometer is the instrument that measures rpm.

While experience can be used to gauge rpm, the tachometer is essential if you want to be accurate. As with the ASI, the tach will help determine your flight condition — whether you're cruising, climbing or gliding. It's senseless

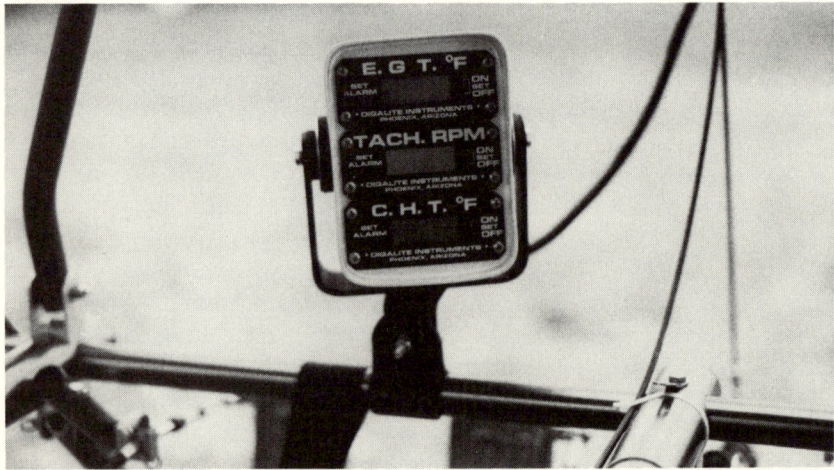

Fig. 11-13. Electronic monitoring of an engine's performance is also available in this self contained, three function unit.

Instrumentation for Ultralights 97

Fig. 11-14. An increasingly popular auxiliary instrument is the Hobbs hour meter, which records the time the engine is running. It is quite useful in rental situations and in keeping track of maintenance.

to fly at wide open throttle, as the engine will wear out all that much sooner. (Cruise speeds are typically quoted for power outputs of from 50% to 75%, resulting in averages of 79% and 94% of top speed. The extra 25% of power for a 6% gain in speed is crazy!)

The tach is also quite useful for helping to determine the engine's state of tune. If you can't rev up to full rated static rpm, something is wrong — don't attempt flight. If you can't hold idle — don't attempt flight. Perhaps the plug is fouling or the needle valves are misadjusting. Maybe you have the wrong prop installed. If you have dual ignition, also short each circuit out when you run-up for takeoff, the rpm should drop for each short. RPM can tell you a lot about your engine — it can tell you not to fly!

A tachometer can be either mechanical or electrical in operation. A mechanial tach contains a governor mechanism which translates centrifugal force into a needle position via levers and gears. The more common electric tach operates from one per rev impulses from the engine which generates an alternating current from the magneto. The current flows to a millivoltmeter and is converted to DC and reads out as rpm.

Cylinder Head Temperature Gauge

Another useful instrument is the cylinder head temperature gauge. The engine manufacturer has determined optimum operating temperatures

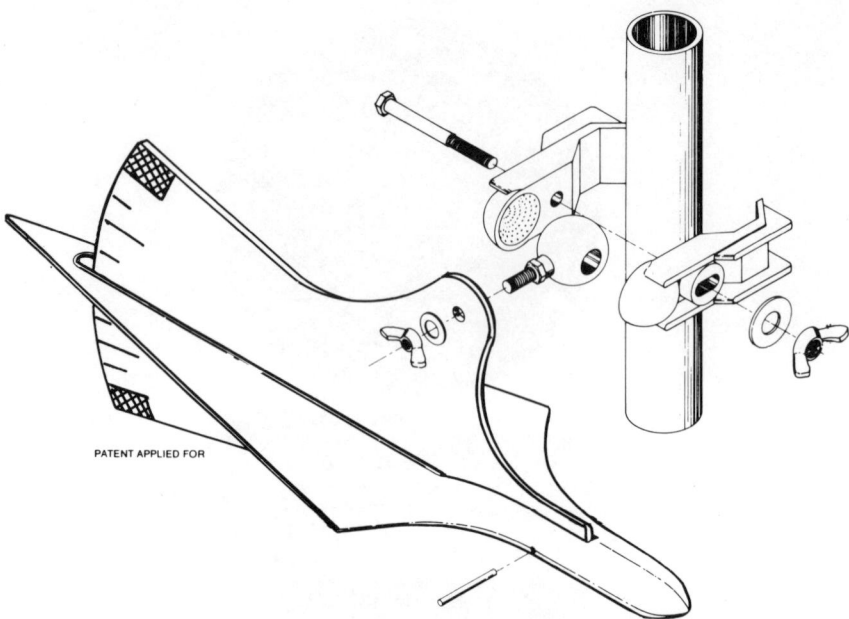

Fig. 11-15. The angle of attack indicator. Markings should correspond to primary flight speeds, such as: best rate of climb, best angle of climb, best glide, maximum duration, etc. Crosshatched area indicates stall.

which, if exceeded, could damage the engine. The CHT will tell you if you're near danger, especially during a prolonged climb or glide. A cool engine could quit easier than a warm one, and they cool down in a glide. Periodic "gunning" of the engine is recommended. It helps keep the engine warm and the plug from fouling.

Too hot an engine can be worse. High temperatures can be caused by too lean a mixture, which implies not enough fuel and oil to dissipate heat and lube the cylinder. The result could be a seizure, requiring an overhaul of the engine. High temperature can also be generated during a long, slow, high throttle climb. Your CHT will tell you what's happening and hopefully, you'll respond accordingly.

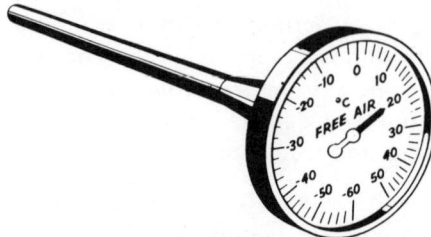

Fig. 11-16. The free air temperature gauge, calibrated in degrees centigrade, can be quite useful in figuring true airspeed and helping determine how you dress for flight.

Instrumentation for Ultralights 99

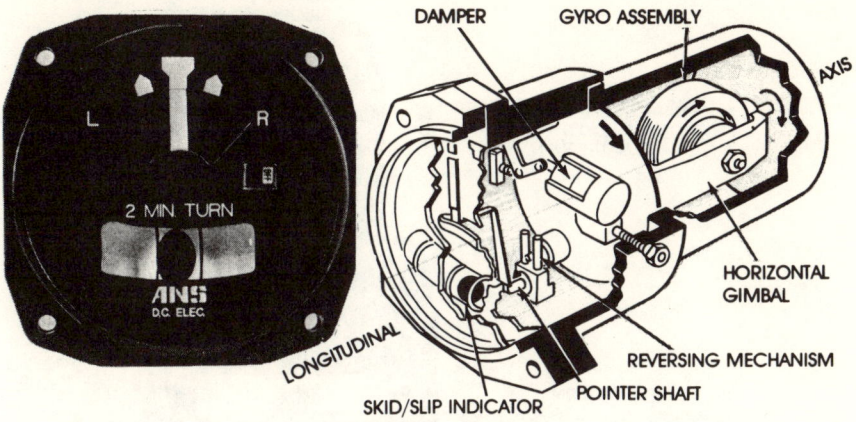

Fig. 11-17. The electric turn and bank indicator and its inner mechanisms. USAF.

Fuel Gauge

A fuel gauge is not necessary, provided your tank is transparent and in view. If the fuel level can't be seen, you'll need a gauge to indicate your fuel level. Remember, out of sight, out of mind. It's embarrassing, at best, to run out of fuel and it can be hazardous to your health. Be aware of your fuel level at all times.

Angle of Attack Indicator

The ultimate device possible for determining your *degreeness of flight* is the angle of attack indicator. Until recently, these have not been commercially available, and the only way to get one was to build it yourself. At any rate, as we've said many times before, the angle of attack is the most

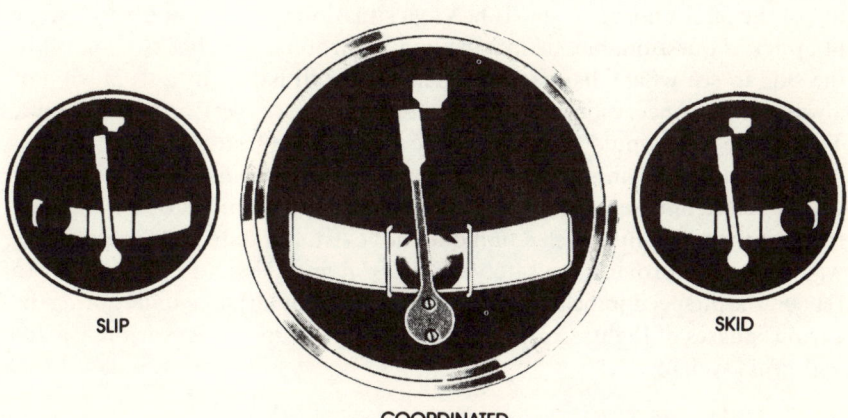

Fig. 11-18., A half standard rate left turn in various degrees of coordination. USAF.

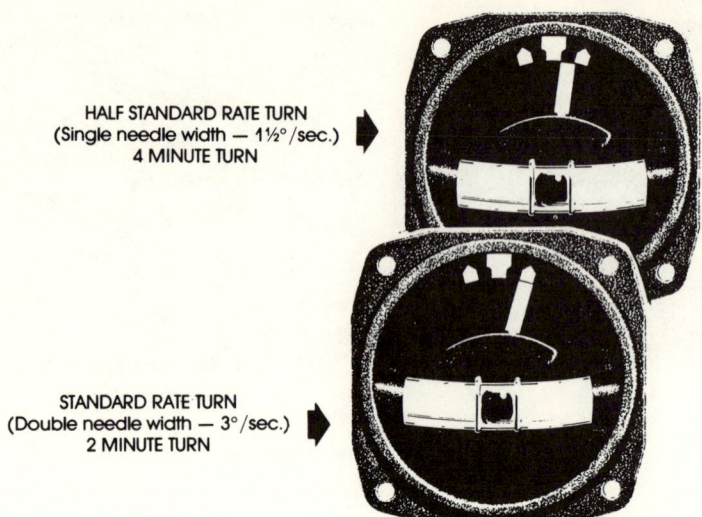

Fig. 11-19. How a coordinated right hand turn is indicated at standard and half standard rate. USAF.

important element of flight. And, no matter what your airspeed, bank angle, or g load, *a wing will always stall at the same angle of attack* — a characteristic of the airplane.

A direct reading angle of attack indicator such as that shown, will work in theory, but has some drawbacks. First of all, it must be mounted at least one chord length in front of the wing and far enough away from the fuselage and propeller slipstream so as not to be affected by the local airflows. In short, it must be in the freestream airflow. If it isn't, it will not give a true reading of the angle of attack. Secondly, it must be well balanced and free moving.

The trouble is, to meet the first requirement, the indicator will have to be out of the pilot's normal view. If he's in a situation where a low enough angle of attack is questionable, he may be inviting trouble if he has to look off to the side to see what's happening. Then too, both halves of a wing will not always stall at once, especially during a turn or while experiencing a side gust. This means you would need two angle of attack indicators to do the job.

The remote viewing problem could be solved by building the indicator with an electronic sensing device connected to an instrument on the panel in front of the pilot. But, such a unit does not exist. If it did, it would go a long way toward improving the safety record of flying. Therefore, we are left to rely on the airspeed indicator and at what speed our airplane stalls during the various phases of flight: straight and level, banked, and under various g loads and gross weights.

12
Density Altitude and the Koch Chart

Density altitude is one subject that nobody ever seems to talk about during those great flying "bull" sessions — probably because many ultralighters simply don't know enough about it! Like most other aspects of flying, being ignorant of the facts can kill you. Knowing your density altitude and what it does to your ultralight's performance is vitally important for every pilot to understand. Hot outside air temperatures, high field elevations, and even high humidity can suddenly change an otherwise normal takeoff or landing into an accident.

The three factors that go into determining density altitude are:
Altitude — Air thins out with altitude.
Temperature — Air thins out with temperature.
Humidity — While humidity is not really considered important in its effect on aerodynamic performance, it does effect the power output of your engine. At high outside air temperatures, the atmosphere can hold a large amount of water vapor. For instance, at 96°F the air's water vapor content can be eight times as great as it is at 42°F! High humidity and high density altitude don't

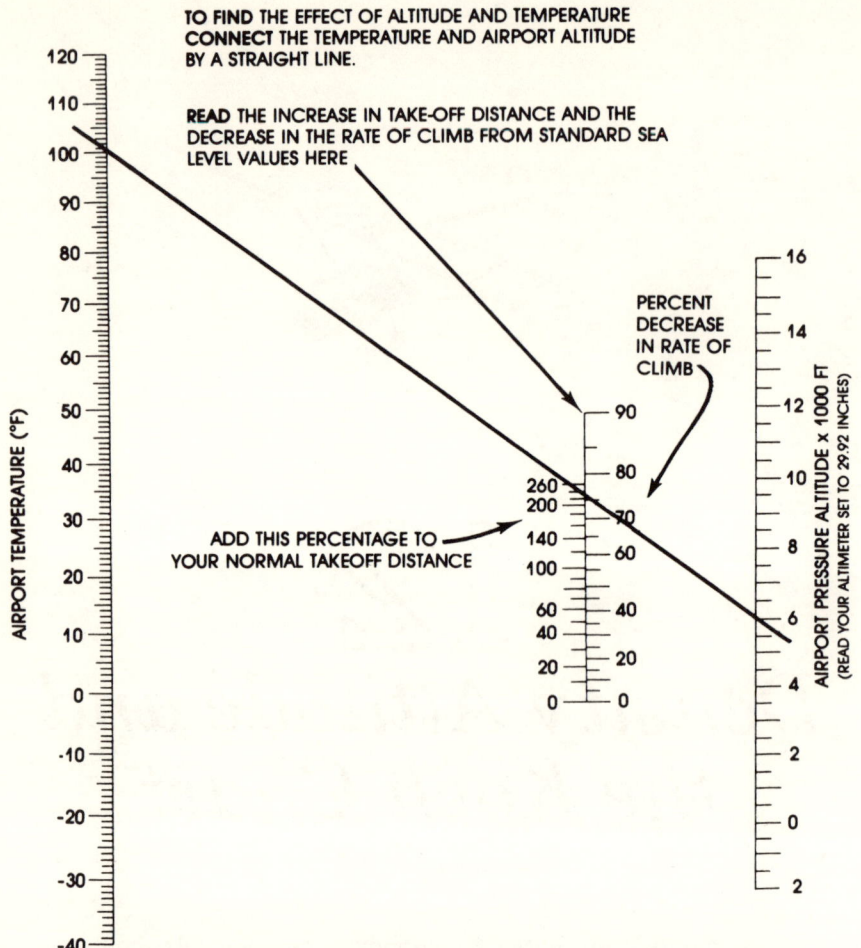

Fig. 12-1. The Koch Chart for altitude and temperature effects is of utmost importance when field temperatures and/or altitudes are high.

necessarily go hand-in-hand, but if high humidity does exist, you'd be wise to add 10% to your computed takeoff distance and be prepared for a reduced rate of climb.

Your Owner's Manual, as prepared by the ultralight manufacturer, should provide you with flight performance figures under standard conditions, i.e.,

Density Altitude and the Koch Chart

sea level at 59°F. But, these figures will change for conditions that are different than standard. If you don't allow for changes in density altitude, you could be in for some real surprises, or a disaster, during takeoff and climbout.

The effects of density altitude are not exclusive to mountainous areas. Whenever the field elevation is above sea level and the temperature is above 59°F, density altitude increases. The main thing about the mountains is that the effects of density altitude are all the more pronounced. Takeoff distances, engine power, as well as climb rate are all adversely affected. The true airspeed increases, while indicated values remain the same.

Whenever you are at density altitudes of 5000 feet or more, or at 75% throttle setting, it is imperative that your engine be leaned properly for maximum power — you need all you can get. Too rich a mixture setting would be detrimental to overall performance, and could lead to plug fouling.

Keep in mind that density altitude is not to be used as a height reference, and it should not be confused with indicated altitude, pressure altitude or true altitude — it is used only in determining ultralight performance capabilities.

Whenever the outside air temperature rises above the standard value of 59°F, the density of the air decreases and the density altitude increases. This causes the engine's maximum power output to be reduced and affects the aircraft's aerodynamic performance. If the air temperature is above standard, no matter what the field elevation is, you should make it a habit of checking your aircraft performance figures and the Koch Chart. The following chart shows "rule-of-thumb" effects of temperature and density altitude:

STD TEMP	ELEV/TEMP	80°F	90°F	100°F	110°F
59°F	Sea Level	1,200'	1,900'	2,500'	3,200'
52°F	2,000'	3,800'	4,400'	5,000'	5,600'
45°F	4,000'	6,300'	6,900'	7,500'	8,100'
38°F	6,000'	8,600'	9,200'	9,800'	10,400'
31°F	8,000'	11,100'	11,700'	12,300'	12,800'

As far as the ultralight pilot is concerned then, increased density altitude means:

1. **Longer takeoff distances.**
2. **A reduced rate of climb.**
3. **The true airspeed will increase (indicated airspeed remains the same, however).**
4. **Increased landing roll.**

Often, at fields located at the higher elevations, like those in the Western United States, flight operations can be affected by density altitude to the extent that it's just about impossible to fly. What this means is that during midday, operations could become quite hazardous — you'd better plan on

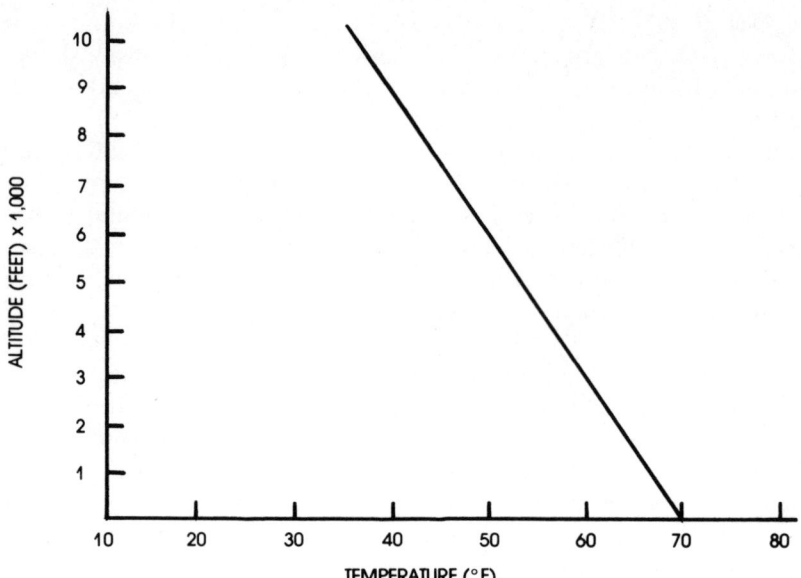

Fig. 12-2. The standard lapse rate has the ambient temperature drop an average 3½°F for every 1,000 feet. This is important to bear in mind when planning your cruising altitude.

flying during the morning or early evening. Interestingly, these times of the day are also more suitable for ultralight operations from the standpoint of wind conditions — it's usually calm in the morning and early evening, while often blustery in the afternoon.

In order to avoid the problems and potential dangers associated with density altitude, you should refer to the performance data presented in your owner's manual as developed by the ultralight manufacturer. If this is not available, the Koch Chart may be used to determine how your ultralight's performance is affected in regards to takeoff distance and rate of climb.

13
The Speeds to Fly an Ultralight

All aircraft have certain speeds at which they should be flown in order to realize a particular level of efficiency, economy, safety or performance. The following speeds could represent an "average" ultralight, based on FAR 103 criteria.

Considering the speeds below as a baseline, the following "rules of thumb" apply. If your airplane has a larger than average wingspan, it will be most efficient in slow flight, and all speeds will decrease. If it has a short span, it is inefficient in slow flight and all speeds should be on the higher side.

If you fly a "dirty" (i.e., cables, struts, floats, etc. in the airstream) ultralight, slower speeds are favored. On the other hand, a "clean" (i.e., streamlined) ultralight will favor slightly higher figures.

Higher wing loadings (e.g., heavier pilots) will cause an upward shift of all speeds, and vice versa.

NOTE: For an in-depth analysis of the various flying speeds of an ultralight, it is suggested you refer to a copy of "Ultralight Flight," UP Book No. 3. To gain a better understanding of kinship of some of these airspeeds, the following excerpt is reproduced here.

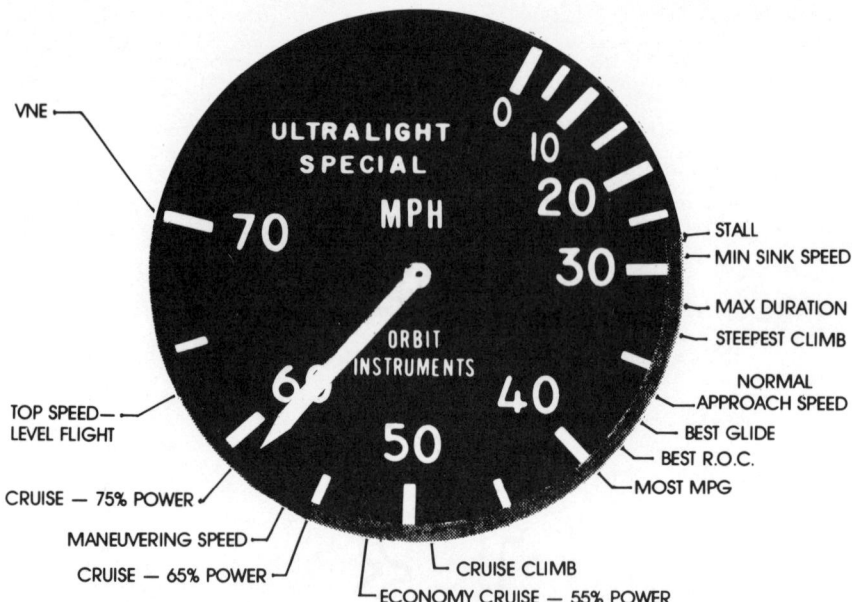

SPEEDS TO FLY TABLE

27.6 MPH - STALLED. Insufficient airspeed (angle of attack) to fly. Ailerons become ineffective while tail controls get sluggish. The nose drops and aircraft falls.
28.9 MPH - MINIMUM SINK speed will allow the aircraft to remain airborne for a maximum period of time while gliding. It is useful for maximizing climb rates under soaring conditions.
32.1 MPH - MAXIMUM DURATION speed will enable the aircraft to remain airborne for the longest period of time while under power. It requires a minimum of the engine's power.
33.1 MPH - BEST ANGLE OF CLIMB speed is where the aircraft climbs at its steepest angle. It is useful for climbing out of short, obstructed fields. Since it is so near the stall, caution should be exercized in its use. It is approximately 1.2 times stall speed.
35.9 MPH - NORMAL APPROACH SPEED is typically set at 1.3 times stall speed.
37.4 MPH - BEST GLIDE speed provides the maximum L/D and the shallowest glide or maximum horizontal distance from a given altitude with the engine off.
38.6 MPH - BEST RATE OF CLIMB (R.O.L.) speed provides a climb to a given altitude in the least amount of time. It is approximately 1.4 times stall speed.
39.9 MPH - MAXIMUM RANGE speed is where you get the most miles per gallon (mpg). It enables you to fly the fartherest on a tank of gas.
48.3 MPH - CRUISE CLIMB speed is somewhere between best R.O.C. and cruising speed. It provides improved engine cooling, a lower nose attitude, and allows you to gain altitude while covering more ground than at the best R.O.C. speed.
51.0 MPH - ECONOMIC CRUISE provides reasonable airspeed with modest fuel consumption. It requires about 55% of the engine's power.
54.5 MPH - CRUISE SPEED at 65% power.
55.2 MPH - MANEUVERING SPEED is traditionally set at twice the aircraft's stall speed, and is supposed to protect against structural failure in turbulence. Here, the aircraft could receive a maximum 4-g load, where it would stall before loading the airframe further. Of course, the aircraft must be capable of handling 4-g's in the first place.
58.0 MPH - CRUISE SPEED at 75% power.
63.3 MPH - TOP SPEED for level flight at full throttle.
70.0 MPH - Vne (VELOCITY NEVER EXCEED), beyond which structural damage will result. Its exact value depends on the particular design.

Fig. 13-1. An ultralight must be flown at a specific airspeed to realize a particular level of performance.

Best Rate of Climb, Range, Endurance and Angle of Climb

When the throttle and trim are set to produce an airspeed slightly above the best glide angle speed, which occurs at around 38.6 mph for the "average" ultralight, the airplane-engine-propeller combination is most efficient. During a power-off glide, 37.4 mph is the most efficient airspeed. But, since propellers are usually designed for cruising speeds, the engine-propeller combination is more efficient at cruise. This means the aircraft under power will have the most excess power available at a speed slightly higher than the best gliding speed. Naturally, the higher the throttle setting, the higher the rate of climb.

Notice that the excess power available doesn't decrease much in a band about 4 mph on either side of maximum, meaning it's not that critical a speed. However, it is better to be on the high side of the "best" speed. It improves engine cooling, and allows for an improved airspeed margin above stall — in case the engine quits.

If you are trying to set a range record, the "best" speed to fly is slightly greater than the best rate of climb speed which is 39.9 mph in this case. If our propeller is designed for climb, the best rate of climb speed might be at or below the best gliding speed. For most practical purposes however, the best rate of climb speed can be used for maximum range and best glide — it all depends on the particular airplane-propeller-engine combination and where it is most efficient.

In order to stay airborne the longest period of time, it is necessary to fly at a speed slightly above the speed for minimum sink. Power-off, minimum sinking speed will yield the longest time aloft from a given altitude. Power on, the engine propeller efficiencies cause an upward shift in this speed. Slightly above maximum duration speed, an airplane will attain its steepest angle of climb, again depending on the engine-propeller combination.

"Maintain thy airspeed, least the ground rise up and smite thee."

14
Seaplane Operations

Float flying is a very exciting facet of ultralight aviation, opening up vast recreational areas to the pilot. This chapter will give you a good knowledge of the basics of operating off water. We'll begin by providing a description and illustration as an aid to recognizing the various parts of floats. These terms are used occasionally in the text and this way you can refer back to the illustration should you be in question about a certain item.

Float Terminology

Slow Taxiing — Moving through the water at a speed which does not allow the float to plane on the surface of the water.

Step Taxiing — Moving through the water at a speed which causes the forebody to plane or ride on the surface of the water with minimum displacement.

Planing — When the float moves at a great enough speed to cause it to ride up on the surface of the water, with the afterbody completely clear.

Suction — The tendency of a float to "stick" to the water during takeoff, due to surface tension and other factors.

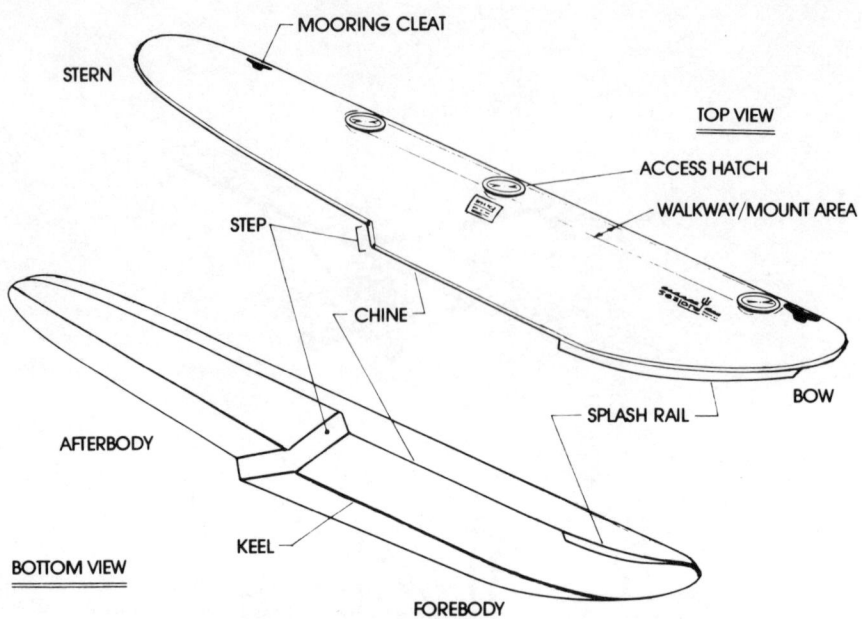

Fig. 14-1. Top and bottom views of a typical ultralight float, including terminology.

Preflight

A preflight check of the float system is as much a mandatory part of seaplane flying as is the preflight of the aircraft itself. Although seaplane ultralight flight is a safe and easy task, one unsafetied pin that becomes dislodged in flight could cause a float to drop out of position and wreak havoc upon landing. The preflight check is your own final assurance of a safe flight.

Check inside floats for possible water accumulation that might slosh to the back, affecting the aircraft CG.

Check that all *mounting bracket bolts* are snug.

Check all support strut *pins, bolts, nuts and safety clips* to assure they are properly secure.

Observe the movement of the floats with the waves when taxiing out for takeoff, as a final check to assure their movement is not in excess of normal, and that they are sound and secure.

If you find a problem, correct it before flying. It's virtually impossible to correct later on when it begins to generate trouble in the air. Check everything thoroughly before you fly.

TAXIING, TAKEOFF AND LANDING

Taxiing, Takeoff and Landing described in this section are assumed to be under normal conditions: Light wind, light chop to the water, no turbulence causing obstructions and direct into the wind conditions. These are probably

Seaplane Operations

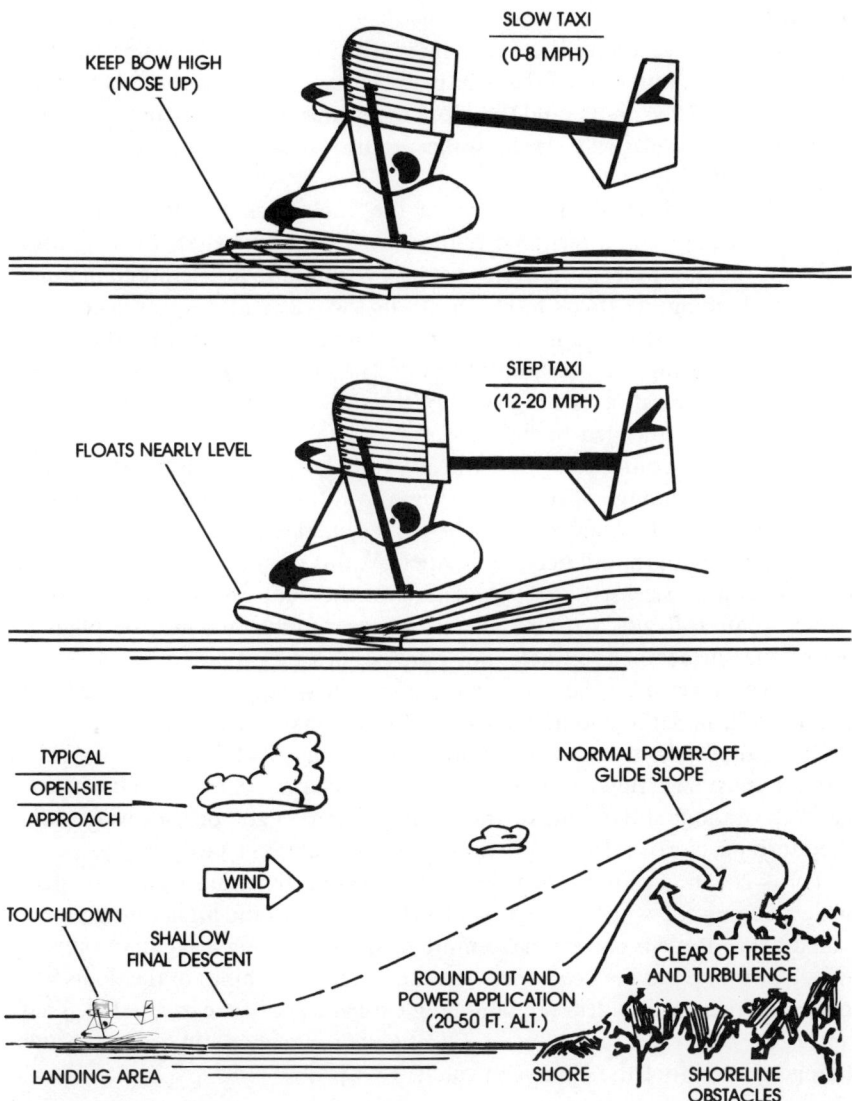

Fig. 14-2. The basics of seaplane operations.

the best circumstances for initial trial flights if you have no prior seaplane experience.

Get a feel for the handling of the aircraft on the water by doing slow taxiing, then step taxiing both upwind and downwind. Progress to short hop flights of about ten to twenty feet of altitude after becoming comfortable with the taxi exercises. When making actual flights, keep in mind the increased airspeeds you should use, the reduced climb rate and the steeper glide slope for approach to landing.

Taxiing

Slow taxiing is accomplished with power settings from idle on up to half or more. Since the hull of the float is displacing the water instead of planing on top, the drag of the floats (and therefore the power required to move them) is fairly high. A relatively large wake is also produced because of the displacement of the water.

Keeping the bow up by the use of the back stick or aft weight shift improves the splash pattern and reduces the drag somewhat. Bow up also quickens the time and shortens the distance required to accelerate to planing speed by helping the floats to climb out of the water at a lower speed.

Slow taxi is best suited for precision maneuvering to and from the dock, and for positioning the aircraft for the initial takeoff run. The power required to propel the aircraft at the high end of the slow taxi speed range is actually greater than that needed to step taxi.

To step taxi, continue to add power while keeping the bow up. As the aircraft attains enough speed it will begin to ride much higher in the water than in previous slow taxi stages. The wake generated by the floats will also change in its pattern and become smaller. At this point, some of the bow up pressure can be released and the attitude of the floats allowed to level out. Now the aircraft will actually accelerate several mph and will be planing across the surface of the water, not plowing through it.

Reduce power after the floats plane out since the reduction in drag from planing will be large enough to let the aircraft take off if this is not done.

Since airspeeds of ten to fifteen mph are common while step taxiing, the aircraft must be "flown" using all the aerodynamic controls. The speeds generate enough lift to make the ultralight very light on the water and therefore gusts can push it sideways or bank the wings fairly easily.

Turns can be accomplished while step taxiing by using rudder to steer while keeping the wings level. If your taxi speed is in the higher range (near takeoff speed) turns can be made more positive by using a slight amount of nose down to place the keels deeper in the water. The chines of the floats will not catch the water unless the bank angle during the turn exceeds 15°. That angle of bank will cause most ultralights to drag a tip in the water (dangerous) before the chine can catch.

Takeoff

For first flights, use a waterway that has about 1000 feet of "runway" and about 2000 feet before any tall (50 foot) obstacles. After becoming familiar with the takeoff and climb performance of your ultralight, these figures may be adjusted. Normally, 400 feet or less will be required to reach liftoff and, after traveling 2000 feet you may have 150 to 300 feet of altitude.

Taxi to the downwind end of the waterway and turn the aircraft into the wind. With power at idle the aircraft will want to weathervane directly into the wind — an excellent substitute for a windsock. Check to assure the takeoff area is clear of boaters, swimmers and fishermen.

Seaplane Operations 113

Fig. 14-3. Quicksilver with amphibias landing gear. Wheels retract for water operation.

Apply full power. The aircraft will plane out on the step and continue to accelerate until barely skipping across the wavetops. Use back stick to rotate and lift the floats out of the water (this is done at the minimum takeoff speed to help reduce water drag at the earliest possible point). Let the aircraft accelerate to the best climb rate airspeed while flying level just above the water and then initiate a normal climbout.

Fly a standard pattern around your "field" one or more times to feel out the new handling of the aircraft. You will probably notice an increased pendulum stability. With enough altitude at hand, check out the feel at minimum control airspeed very carefully. Check the glide slope and attitude at various power settings and airspeeds.

Landing

Set up your standard approach and, when turning onto final, carry a little extra power to compensate for the greater sink rate caused by the floats. As you near the surface of the water, level out but do not flair so as to lose airspeed. The trick is to touch down in a level attitude flying somewhat faster than you would if landing on the ground. This technique yields the smoothest landings. When you have leveled out and are at approximately a foot or two above the surface, smoothly reduce the power to idle and let the aircraft settle on the water.

After touchdown, carry enough power to step taxi back to the dock. Or, as the speed decreases and the floats begin to come down off the step, you can smoothly apply full nose up which will drag the afterbody of the float causing drag in the water and rapid deceleration.

Careful experimentation with the flight techniques described can yield the best operational methods for your ultralight. Take care to approach each new trial with caution until it is proven out.

Special Flight Conditions

Special flight conditions concern realms of operation where performance or handling is affected and different piloting technique is required. Since the conditions covered in this section will occur from time to time in normal operation, it is a safety oriented section and worth the time it takes to read and remember.

Since most waterways are not equipped for aircraft operations like an airport, it helps to learn a few techniques for deriving the needed information from mother nature. Wind direction can easily be determined by observing waterfowl to see which direction they are floating (they always float facing into the wind). Moored boats also weathervane into the wind. Wave direction and calm areas behind tree lines give you an indication of wind direction, too. Wind velocity is also apparent by observing the surface condition of the water — from calm to wind whipped white caps.

Crosswind technique is mostly aimed at ultralights with three axis controls, since ailerons are a necessity in maintaining wings level in a crosswind. If you will be doing crosswind takeoffs and/or landings it is important to be aware of turbulent air that may be blowing off a tree line, etc. upwind of your takeoff area. The rotor effect of the wind over these obstructions can severely limit your climb rate and control in blowing back

Fig. 14-4. Pontoons with retractable gear built-in, offer lower drag.

Seaplane Operations

down to the surface. This effect is especially apparent when flying off a river oriented in a crosswind fashion.

During the takeoff, the important thing is to cross control using the rudder to maintain a straight track across the water while using the roll control to keep the crosswind from lifting a wing. Once airborne, flying the ultralight should be as usual. Landings should be made by crabbing (flying with wings level and nose pointed toward the wind) to maintain a straight track over the landing area. During the level-off just above the water (1-2 feet of altitude), use the rudder to line the plane up with the "runway" and let it settle in immediately. Continue to steer with the rudder until slowed. This maneuver is necessary to prevent touchdown with the aircraft crabbing or drifting crosswind, which could result in a tip over.

Takeoff from glassy-smooth water will require special techniques to break loose the suction and lift off. This is because the lack of aeration under the floats increases the surface tension effects of the water. Without special takeoff techniques the takeoff run would consume long distances in order to generate enough lift to literally yank the floats from the water.

The first method is to taxi around in the water to make your own waves from the wake of the floats. Timing should be such that you ride over your own wake during the planing phase of takeoff, thereby introducing air under the floats at about the normal takeoff point.

The second method is to use your roll control (ailerons or spoilerons) to lift one float out of the water at a time. This way you are only breaking one-half of the total suction at a time making the liftoff much easier. Take care not to bank the aircraft too steeply — only a few degrees of bank are required to lift a float out.

The third method is to "horse" the aircraft fore and aft, using a pumping motion with the joystick. When done with the correct timing, the floats will alternately sink in and spring out of the water until you reach a point when the airspeed and "spring out" will allow you to yank the aircraft out of the water. Once airborne, let the aircraft accelerate to climb speed and continue on in a normal manner.

Usually, a combination of these smooth water methods will be used together. Landing on smooth water doesn't usually present a problem except for height judgement over the water.

When landing out in the open in large bodies of water or when it is nearly dark, altitude judgement becomes more difficult. In twilight conditions, landing nearer to shore can provide you with an altitude reference. Out in open waters, a slow letdown until contact is best since you probably have virtually unlimited space anyway.

Landing in open seas with large swells requires a different approach than the typical landing under inland conditions. Takeoffs during these conditions requires similar techniques. Runs should be made parallel to the waves in the trough between the waves to avoid premature liftoff. This will usually result in a crosswing takeoff or landing, so be cautioned to use the appropriate crosswind technique. If the swells are fairly close together it is essential that you should land parallel to them so that the bow of the float does not dangerously bury itself by piercing into the wave.

When landing in large swell conditions, pay particular attention to the wind direction — on shore or off shore. This is of concern especially in relation to the size of the body of water you're flying from (the Great Lakes or the Oceans) and your engine's reliability.

Deadstick (power-out) landings should be executed with regard for the increased sink rate and steeper glide slope of the float equipped ultralight. Keep your approach altitude relatively high, using shallow dives to dissipate any excess. The airspeed for flair should be fast so a smooth, level attitude touchdown can be made as with power-on landings. Avoid making deadstick downwind landings if at all possible, especially in stronger winds.

If the need arises to make an emergency ground landing, the following technique will result in the safest landing with minimal or no damage to the floats. A normal approach is made and when making the final flair, airspeed is bled off with a nose-slightly-high attitude just above the ground. Do not slow the aircraft completely to stall speed as this results in the nose being too high. The aircraft will settle to the ground when airspeed finally decays and slide to a stop in about thirty feet. Land into the wind if possible to keep the ground speed at a minimum. Of course, try to land on grass, or the slipperiest surface possible.

Seaplane Safety Considerations

Here's a compilation of things to know or watch for during seaplane operation.

Sea breezes on the Oceans or Great Lakes shores will change directions, from on-shore to off-shore, vice versa, or become completely calm. These changes occur near dawn or sunset, plus or minus a few hours. Watch for tell-tale signs of these shifts.

Watch carefully for *power lines* that are often strung across rivers and narrow lakes at low heights.

Watch for *boaters, skiers, fishermen and rubberneckers* that may dart into your takeoff or landing path without warning. Also watch for *swimmers* which are sometimes hard to see. In navigable waterways, *buoys* are another hazard to watch for when flying low or landing.

Submerged rocks, stumps, etc. just below the surface present a hazard and should be checked for while still in the air in the landing pattern.

Check *local ordinances* before flying the first time from resort lakes or lakes within a city limit. Sometimes they may have an ordinance against seaplane operations, and it is best to find this out, before receiving a citation.

It is common sense to *carry tiedown line and tools* in the floats to provide mooring and repair capability if you should become *stranded* out in the water or on a river while flying. Carrying some money for food or phone might not be a bad idea either.

Be safe and let someone else know where you'll be flying, especially on longer trips, just in case you have trouble. A *prearranged check-in* time is a good, safe practice.

Use caution when flying in areas with large flocks of *waterfowl.* A bird strike in the prop could be dangerous if the engine is not shut down immediately to stop the vibration.

When operating in *beach areas,* watch for *other aircraft traffic* flying parallel to the beach, possibly in both directions.

Always wear a *floatation jacket* or some type of life preserver, and wear *protective headgear.*

Note: When operating from a river, it is often advantageous to take off with the current, even if it's with the wind. The current will aid your acceleration.

For further information on float flying contact Composite Industries, Flight Systems Division as listed in the Appendix.

"It must be understood that ground speed has nothing to do with the action of the ultralight in the air."

15
Snow Ski Operations

If, like most of us, you do your flying outside of the sunbelt you can't help but dread the coming of winter. It starts with chilly morning flying sessions and, as the trees don their fall colors, you soon begin giving up your dawn patrols, concentrating on the warmer afternoons. Inevitably it begins getting uncomfortably cold even late in the day, at which point you may resign yourself to packing up your ultralight with nothing to look forward to but a winter of hanger flying in the company of other ultralight pilots who are suffering the same extended withdrawal symptoms. Well, there is some good news for you adventurous types - you CAN fly your ultralight for twelve months a year without moving to a warmer climate! Winter flying has its own attractions that make it more worth the effort as long as you follow some basic guidelines appropriate to the area in which you live. The whole purpose of this chapter is to open up that whole new flying season and, we think you will find it is well worth the effort!

THE LIMITS OF WINTER FLYING

If you think this little section is about cold weather you are mistaken. Cold weather by itself is no barrier to enjoying your ultralight. Within reason you can expect to be able to fly in comfort over the typical range of temperatures encountered in the northern tier of states. The real limitation to your winter fun has to do with the geometry of the solar system! Winter days are short days and that is where you are going to encounter frustration. A typical summer flying season will see you out taking a spin before going to work with plenty of flying time in the evening after supper! Not so in the winter. Those nasty short days will find you driving to and from work in the dark and weekday flying just isn't in the cards. If you are pretty much confined to the weekends, you will soon discover that the odds will be that you tend to get weathered out with frustrating frequency. It is not that winter weather is particularly bad — it can in fact be more stable than summer — but if you are confined to flying only two days out of every seven the odds will take their toll. The only point here is to make you aware of the fact that you simply can't get as much flying in during the winter months. Fortunately, the winter flying experience is a unique one so, the time you do get in will be of high quality. It sure beats talking about what you will be doing next summer!

There is, obviously, a full range of winter flying conditions, depending upon where you are flying. In many parts of the south we are simply talking about chilly weather in the 20-50 degree range. Snow is rare in such areas so winter flying is simply a matter of dressing for the nippy weather. Farther north we have to deal with zero-degree flying and the presence of snow opens up the possibility of operating with skis — a new world indeed! Most of what follows will be slanted toward cold and snowy winters. Obviously, if your conditions are not that extreme, you will not need all the hints and techniques that are required for Michigan and Minnesota! None of that really matters — the name of the game is winter flying so we can all keep up with the sunbelt types! Let's look at how it is done and why it is worth all the effort!

CLOTHING

This is the real key to winter operations. You cannot fly in comfort, let alone safety, if you are slowly freezing to death. A debilitating chill can come quickly when you are sitting in a steady 30-40 mile breeze. Modern technology has provided some very effective (and expensive) exposure suits and you can even go to electrically heated suits and gloves if money and weight are no object. Such extremes are not required however, for if you live in an area with cold winter weather you probably have all the essential ingredients of your arctic flying outfit. It is simply a matter of fitting things together in the right combination to be effective in keeping you comfortable in the air.

The key is the use of multiple insulating layers to retain body heat. Your level of physical activity will be low and this, coupled with the constant wind-chill, places some real demands on your system, if you don't keep the wind out and the heat in! The basic ingredient is a good suit of thermal underwear. Although wool slacks would be ideal, any close-weave fabric (such as new denim) will do for leg protection to back up the thermal underwear. In really cold weather, additional upper body protection is often required and this can start with an old sweater directly over the underwear, followed by a wool shirt. A relatively light nylon jacket over the shirt, followed by a heavier insulated jacket, will usually complete the basic flying gear. A wool scarf, carefully tucked in to avoid fouling things like props and wires, will seal off your neck.

Feet require their usual measure of attention in cold weather. Two pairs of wool socks, followed by a pair of lace-up leather boots will usually do the job in this department. The pants should be tucked into the outer pair of socks to keep the wind out. The boots should have soles with good traction since you will probably be slogging around in the snow and, we don't want feet slipping from footrests or rudder pedals.

Most fliers find that their hands present the biggest problem. Conventional gloves rarely work in really cold weather, what with the slipstream and frigid aluminum tubes! Down insulated snowmobile mitts provide a good tradeoff in flexibility and feel and will do an excellent job of keeping your hands functional during long flights. Mitts insulated with artificial fibers tend to be colder and are definitely on the bulky side for maintaining a good feel for the controls.

The final details concern face and eye protection. Frostbite is something that can creep up without warning, so full-face protection is definitely a good idea. A pull-over wool ski mask will avoid the problem and keep you comfortable in the coldest weather. You have two options for eye protection — a face shield or ski goggles. Except for really mild winter weather, you should avoid the face shield as it can quickly fog up (or frost over) even if you use some of the fancy anti-fog sprays. I can remember an interesting (!) flight where a face shield froze up immediately after liftoff, requiring that the early part of the pattern be flown by peering out from underneath the shield! Once I got to altitude I had to pull the shield and stow it for the rest of the flight and that is an additional complication anyone can do without! Ventilated ski goggles will not fog up and they are available with replaceable filter lenses that can reduce eyestrain due to snow-glare.

All of this costuming will definitely up your flying weight but, will present few problems given the added performance your ultralight will deliver in cold weather. If your aircraft has acceptable summer performance you can expect to come out ahead in the winter, even with the added weight you will be carrying.

EQUIPMENT CONSIDERATIONS

Winter flying immediately brings to mind the subject of skis. That is something we will cover a bit later. For now, lets look at some basic aircraft maintenance and operating techniques that apply to winter conditions.

Pre-Season Inspection

At some point in the fall you should break your aircraft down and inspect all parts for wear and corrosion. The realities of winter weather will make it harder to do such a task so the end of the usually intense summer flying season is a good time for the chore and will assure that you start the winter in the best possible shape. Complete airframe inspection and parts replacement can then be repeated in the spring to get ready for the upcoming summer.

At this point, you should also give some thought to storage. Generally, it is best to keep an ultralight assembled for winter flying, if only because breakdown and setup are harder to do properly in very cold weather. If you keep the plane assembled however, it should be stored under cover, be it a barn, hanger, or whatever. Winter storms can be severe and you don't need the added burden of snow cover to stress flying wires and support members. If you must break down your ship with each session be extremely careful about setup and rigging since conditions will be far from ideal. It is a good idea to keep pip-pins and other fasteners in a dry storage area so there is little chance of water freezing internally in the pins, impeding their function. The basic materials of most ultralights are certainly stable well below zero but aluminum and some plastics can become brittle at extremely low temperatures. If you plan to fly in temperatures lower than 20 below, you should probably check with the manufacturer, unless you are really familiar with the low temperature characteristics of all the materials.

Preflight

Preflight-preflight-preflight! It seems like people who write books and articles on ultralights have a one track mind. We could point out that those who accumulate enough experience to write all that stuff are the ones who DO their preflight but, there is a special reason for bringing it up here. You are going to be trudging around your ultralight — maybe in deep snow — and it will be cold. You will almost always feel colder on the ground since you don't have that ultralight high to mask out the physical reality of the weather. The fact is that you are not ready either physically or psychologically to do the same thorough job of preflighting that you did during the summer. The only way out of that potentially hazardous box is to take special pains to do the preflight properly! Have a thermos of hot coffee on hand, take breaks as needed but, check the ship out! Pay particular attention to the possibility of cracks in fatigued metal or plastic at low temperatures. Rubber drive belts will be less flexible and should be checked carefully for tracking and the possibility of cracks. Fittings down near the landing gear or skis are likely to be crusted with snow or ice so look twice, cleaning the fitting to make sure all is well. Retaining pins are easily snagged in icy snow (as they can be in grass) so check them with particular care. The key is to make sure you are taking the time to really see everything!

Engines

In thinking about winter, you are probably worried about starting your engine. Take heart. Two-cycle engines really love cold weather! If this were not the case we wouldn't have snowblowers or snowmobiles. A well-

Snow Ski Operations

Fig. 15-1. Preflighting your skis is most important to safe operations.

maintained engine will start easily if you choke it. You will want to hold the choke for the first 15-30 seconds to give the engine a chance to heat up a bit for smooth running. The engine can be allowed to warm up at fast-idle as there is little chance of overheating it in typical winter weather. If you have a cylinder head temperature gauge you should let the engine get up to at least 250 degrees (F.) before loading it up. Engine runups should be done regularly and a tach is very useful to check top-end performance. You won't be flying as often in the winter and your perceptions of how the engine should feel or sound will not be as sharp as they were during the summer. Rings may be a little sticky after extended cold storage so you may not develop full power initially. Hold off flying. Cycle the throttle and the condition should clear up as the engine gets up to temperature. If it doesn't, you should break down the engine to find the trouble.

A few additional engine tips are in order. The first concerns oil. A quality two-cycle oil will mix and stay mixed with gas at any temperature. Some bargain oils may not be so versatile, so it pays to be careful with oil selection. Sparkplugs are another item that may require attention in really cold weather. Your engine naturally cools down at low throttle and, this is especially true in cold weather. Normally you need to be alert to cycle the throttle every once in a while to avoid the problem of loading up the plug, which would result in engine failure. This problem can be more acute in the winter due to the lower ambient temperatures. You may want to go one step hotter in plug heat range (relative to the engine manufacturers recommendations) for winter flying. With a slightly hotter plug the same attention that you give the throttle in the summer should be for winter conditions. BE VERY CAREFUL to change the hot plug back to the standard value prior to the advent of warm weather. Failure to observe this precaution can result in pre-ignition in the hot weather with a loss of power when you need it most and, the real possibility of engine damage.

Overall, you will be extremely pleased with cold weather engine performance. The cold, dense air results in higher power output and less chance of overheating. These two factors, combined with increased prop thrust in the cold air result in a major performance increase for any ultralight.

WINTER FLYING

Mention has already been made about increased engine performance. In addition to an added boost from your power plant, you will almost feel like you have traded up to an airborne hotrod. The denser air of winter will give you a density altitude lower than you have ever experienced and the aircraft will respond with greater lift and noticeably improved maneuverability. The added lift will result in shorter takeoff runs, faster climbout, and increased glide. The stall speed will also be noticeably reduced. The fun thing is that you get all of this with only one tradeoff — you have to dress for the weather!

All of these performance pluses make for safer flying. Provided it is not too cold for comfortable flying, winter is an excellent time to train students. The aircraft will have a far more "solid" feel than in the thinner air of summer. Beginner and experienced fliers will appreciate the stability that is typical of winter air and much of the "rocky road" turbulence of a summer afternoon or early evening is completely lacking. This does not mean there is no thermal activity! Differential ground heating will occur on any sunny day but, the convection cells are larger and the areas of lift are more extended and uniform. You can get some spectacular thermal flights without the real beating characteristic of summer flying.

If it sounds like winter flying is fun, you have just stumbled on the secret of why we do it! There is one other secret to reveal however — sort of the icing on the cake. That secret involves the real pleasures of skis — the subject of this next section.

SKI OPERATION
Ski Design

There is nothing particularly magic about ski design and most major manufacturers have skis that are recommended for their particular aircraft. They can be made of metal or plastic but most are laminated wood. Usually they are simple, flat-bottomed designs, turned up at the tip, with no fancy edges or other gimmicks. Wood skis will typically be laminated with a bonded top and bottom surface of some sort of polymer compound. Virtually all are designed with mounting stanchions that simply slide onto the axles in place of the wheels. A tail-dragger ultralight will have two mains

Snow Ski Operations

Fig. 15-2. Close-up of properly rigged skis.

skis and usually a small one to replace the tail wheel. Trike-gear ships will have two mains and a nose ski.

The skis must have the ability to rotate to some degree to follow irregularities on the snow surface but they must also be held in a distinctly tip-high attitude in flight. This to assure you do not dig in the tips if your landing approach is a bit too steep! In many designs, the rear of the ski is restrained with a small guy wire while the front is loaded with a bungee chord. This holds the ski at a specific attitude with no weight applied but permits rotation when taxiing on an uneven surface.

Manufacturers recommendations regarding the proper assembly of the skis and their associated hardware, the locations of tangs and other attachment hardware, and any other aspects of installation should be followed to the letter. Other than the skis themselves, most of the attachment hardware can be left in place year-round so that switching from skis to wheels after your initial installation is quite easy.

One item not covered in most installation instructions is the care and feeding of the bungee chords that hold the skis in the proper flight attitude. Experienced fliers will not store the machine with the bungees in place. Rubber loses a great deal of resiliency in the cold and the bungee can often lose much of its effectiveness if stored in a stretched condition. At a minimum, you should unhook the bungees when storing the plane. Complete removal and storage in a warmer area is ideal. Hooking a piece of bright ribbon or other material to the bungee attachment points can serve as a reminder to hook them in during preflight! The chords should also be safetied since the failure of a bungee in flight can leave a ski in a dangerous tip-down condition. Rotate the aircraft to get the skis just slightly nose high (bungees in place) and install a safety wire that will allow the tip to drop no further. In this way, should a bungee fail, the ski will maintain a safe orientation for landing.

Ground Handling

It is a fact of life that ultralights were designed to fly and most simply are as clumsy as a gooney bird when it comes to ground handling. Not so when the same ship is equipped with skis! Ground handling will be essentially effortless and you will have all of the maneuverability and speed of a powered ice boat. Excellent low-speed ground handling is the hallmark of ski operations, and you will positively enjoy getting the ship out to the takeoff area. Stationary in-place turns are hard to pull off since you can't get traction in the snow but otherwise you will find it extremely easy to cruise anywhere in the field you wish.

Some preliminary ground handling is suggested before any flying session. Snow conditions can vary markedly from day to day and a few runs around the field area will give you the feel for the current conditions. This also serves to clean off ice or other crud that may have frozen to the lower ski surface while the ultralight was stored. Since you have to warm up the engine anyway this is a good way to check it out. Besides, it's just plain fun! If you are flying from portions of the field you do not normally use — a common situation since any snow-covered open area is suitable for use — you should take time to check for obstructions like stumps, fenceposts, and other nasties that may be buried right below the snow surface. It is far better to spot such items at moderate taxi speeds rather than during a full-throttle takeoff run or while on final. In the case of really deep snow, fences with wire strands at or just below the snow surface can be a class A hazard so check your area with extreme care. Once you have completed your cruising around, it is time to get to the business of flying!

Takeoff

Takeoff on skis is a real thrill. Provided the field is large enough in all directions and free of obstructions, takeoff direction is kind of irrelevant unless the traffic makes a specific pattern necessary. Simply point her into the wind and go for it! In this respect skis have much of the flexibility of floats. Two things will be evident right away. The first is the smoothness of the run, particularly if you are used to operating from rough fields.

The second point concerns the feel of the aircraft as you transition to flying speed. On wheels most ultralights require large control inputs at low speed. Less input is required as the ship approaches flying speed and the wing begins to take some of the weight off the wheels and the control surfaces become more effective. You will almost always reach a point in your takeoff run on wheels where you tend to overcontrol slightly as you make the transition,

Snow Ski Operations

Fig. 15-3. The ultralight accelerates quickly during its ski takeoff run.

lagging slightly behind the changing response of the aircraft. On skis, even a relatively inexperienced pilot will avoid this momentary tendency toward overcontrol. In part, this seems to be related to the smooth and effective low speed handling provided by the skis. Even at low speed the aircraft is in a responsive "flight" mode so there is no major transition and hence little chance of lagging behind the aircraft. Students often find that the crow-hop phase of training proceeds much faster on skis and this may be due to the smoother handling response curve.

Once you reach flying speed, the transition to actual flight is so smooth you will almost fail to notice it! Take the time to measure your tracks sometime and you will be amazed at how quickly you get off and climb out. If your plane seems marginal in the summer you will be pleasantly surprised at its winter performance.

In-Flight Maneuvers

The presence of the skis should not alter your trim in any noticeable fashion unless you happen to have extremely heavy wheels! You will note a very solid feel to the aircraft — a consequence of the dense air and extra "bite" of all the controls. This solid feel, coupled with the smoothness of winter air, makes for some excellent flying!

One thing you will notice is the completely different appearance of the landscape with its blanket of snow. It is easy to get disoriented the first few times until you are familiar with the altered view. At other seasons of the year you have been unconsciously using the color and texture of fields as part of your information input, but all that is gone with the even blanket of white!

Once you start tooling around your local area you will begin to notice another plus about this ski business. All those fields that were previously covered with nasty crops are now smooth and inviting. Your options for off-field landings are expanded tremendously and, that ought to relax you a bit. We all know we can land in a corn crop if we have to but it certainly is pleasant to see all that crop — not to mention furrows and other nasties — buried in the snow!

A little experience will show you that a ski-equipped ultralight will barely mark the surface of a snow-covered field and under such conditions many farmers will welcome a visit. It breaks the monotony of winters days, and you may even get some coffee and donuts. Obviously, you should check things out in advance but, in all probability you will be flying in and out of fields that were only tantalizing patches of color all summer long!

This business of off-field operations brings up an interesting point. If you operate from an airport with plowed runways you may be tempted to skip the added expense of skis. This is not really a good idea for if you are forced to make an off-field landing you can certainly get the ship in OK but you will not be flying it out again with wheels! Dragging an ultralight out of a deeply snow-covered field is second only to exiting a muck farm. If your territory is snow covered, you should have skis. This may require that you operate from unplowed sections of the field (obviously with the OK of the field authorities) but is is a small price to pay for the ability to get back out of fields that misfortune has dropped you into!

Landings

Landings are fairly routine, except for a few cautionary notes. If the field is unfamiliar you should check it out for any possible partially buried obstructions. In any case, stay well clear of fence lines in planning the approach. If it looks like a fence should be there but you cannot see it, assume it is buried and avoid the field margin!

As you make your approach you will notice that it is very hard to tell your precise altitude. Normally you are using the texture and details of the field surface in gauging your terminal approach and flare but these cues are lacking on a snow-covered field. You have to develop the habits of float fliers landing on smooth water — use the horizon as your reference — not the poorly defined surface!

Fortunately the dense air will let you do a class job of modulating your flare which is fine if you err on the high side. On the other hand you might misjudge on the low side — in which case you will appreciate the soft, resilient quality of snow! The snow will go a long way toward compensating for your initial errors and, it will not be long before you have the hang of landing on a featureless snow surface. Slideouts should be nice and smooth after touch-down and in moments you will be heading back to your operations area.

Conclusion

Although this has only been a brief introduction to winter flying we hope that it is sufficient to prompt you to give it a try. The general flying conditions and aircraft performance make it well worth the effort. Everyone likes the shirtsleeve flying of summer but the next time you are fighting your way through some sink over a nasty cornfield some July evening you will definitely be thinking of a white Christmas!

16
Basic Navigational Methods and Procedures

If you are not content with flying around within sight of your field and wish to go some distance away, you'll need to know the basics of aerial navigation. The two methods useful in an ultralight are pilotage and dead reckoning, which are explained below.

Charts

For the purposes of navigation, the continental U.S. is divided into 37 sections. The charts used are called Sectionals, which are drawn on a scale of 1/500,000, or about 8 miles to the inch. Each chart contains two sides, one for the northern half and one for the southern half of the section. Sectionals are designed mainly for private cross-country flying and are ideal for pilotage in an ultralight. They contain practically everything you need to know about what you're flying over. Obstructions such as TV towers, tanks and

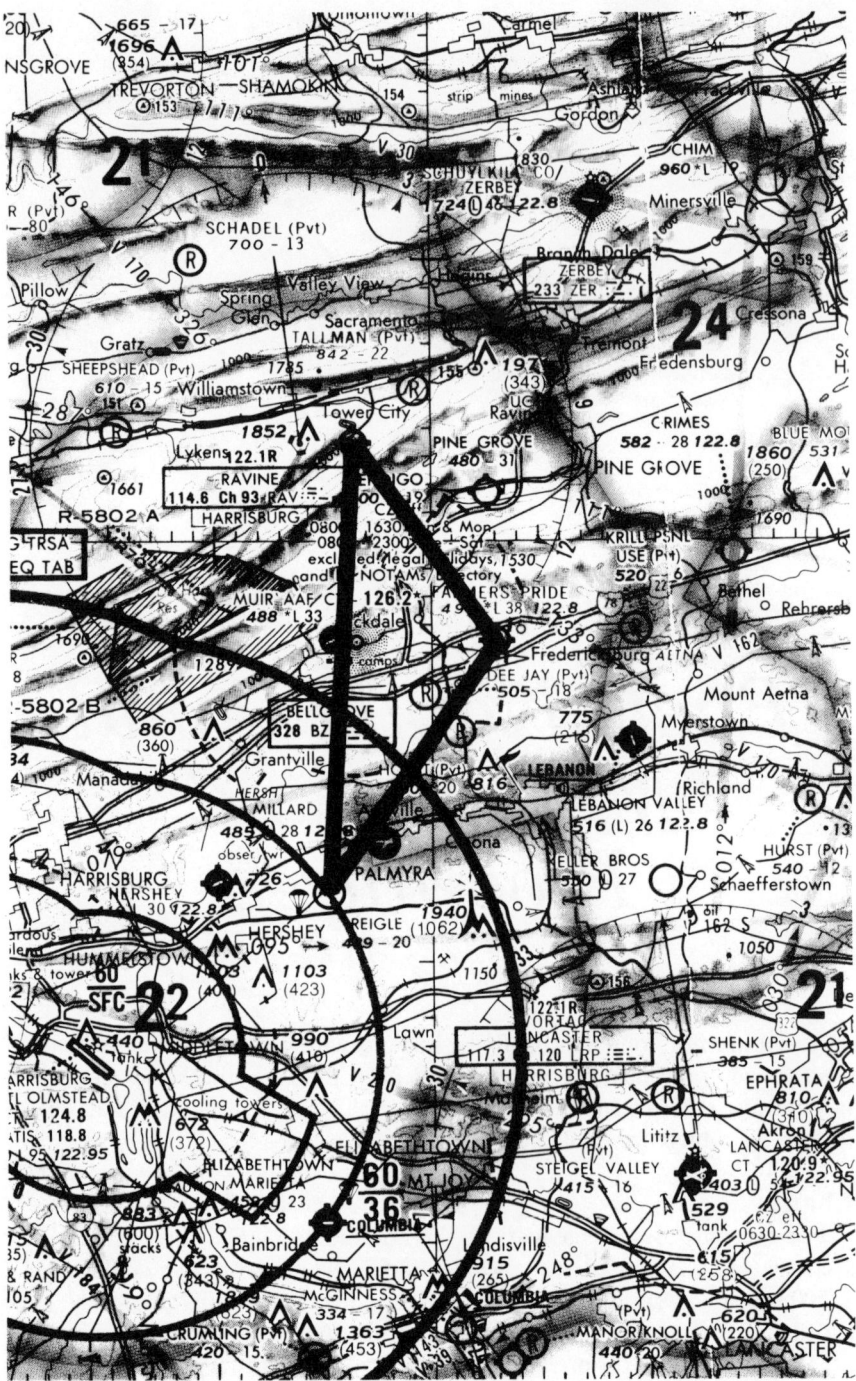

Fig. 16-1. Triangular cross-country flight plotted on the New York Sectional, as described in text.

Basic Navigational Methods and Procedures 131

mountains are shown along with their elevations above sea level. Topographical information, like race tracks, mines and quarries, bridges and viaducts, power transmission lines (though not all), outdoor theatres, piers, cities and towns, dams and roads. Of course, complete information is also presented for all airports, private and commercial, paved and unpaved, controlled and non-controlled. Study and become thoroughly familiar with your Sectional before attempting any cross-country flight.

Pilotage

Pilotage is navigating by means of visual reference to landmarks with the help of a Sectional chart or road map. Landmarks or checkpoints could be a lake, mountain, race track, road intersection, dam or town, etc., and pilotage would involve flying from one to the other, until your flight was completed. You could also follow railroad tracks, roads, transmission lines, mountain ridges, or rivers.

Plan Your Flight

All good pilots are interested in accuracy and safety and the best way to maximize these two is by planning your flight *before* you get into the airplane. You can jump into an ultralight and fly around locally without too much thought, but going cross-country requires proper pre-flight planning. Presented below, is a logical sequence of events in planning your pilotage flight.

Winds and Weather

Before any flight is attempted, wind and weather conditions must be known. Local radio and TV forecasts are fairly accurate (75%) and should tell you what you need to know. Basically, what you want is fair weather, with calm to light winds. And remember, the higher you climb, the faster the wind speed, which also might have a different direction than the surface wind. If you need more detailed information, contact your local Flight Service Station (see your telephone book). Be aware that Flight Service gives wind speed in knots. Convert to mph by dividing knots by 1.15 (11.5 mph = 10 knots).

The Sectional Chart

If you don't already have one, you'll need the latest Sectional covering the area of your intended flight. While they were designed for easy reference at 5,000', they are still useful to ultralights.

Plot Your Course

First obtain a navigational plotter. Open the chart and spread it out on a table. Lay the plotter on the chart and draw a line from your origin to your destination. Next, mark off a small line every 5 or 10 miles. Look at the route and see it if passes over any hazardous terrain or restricted areas — such as

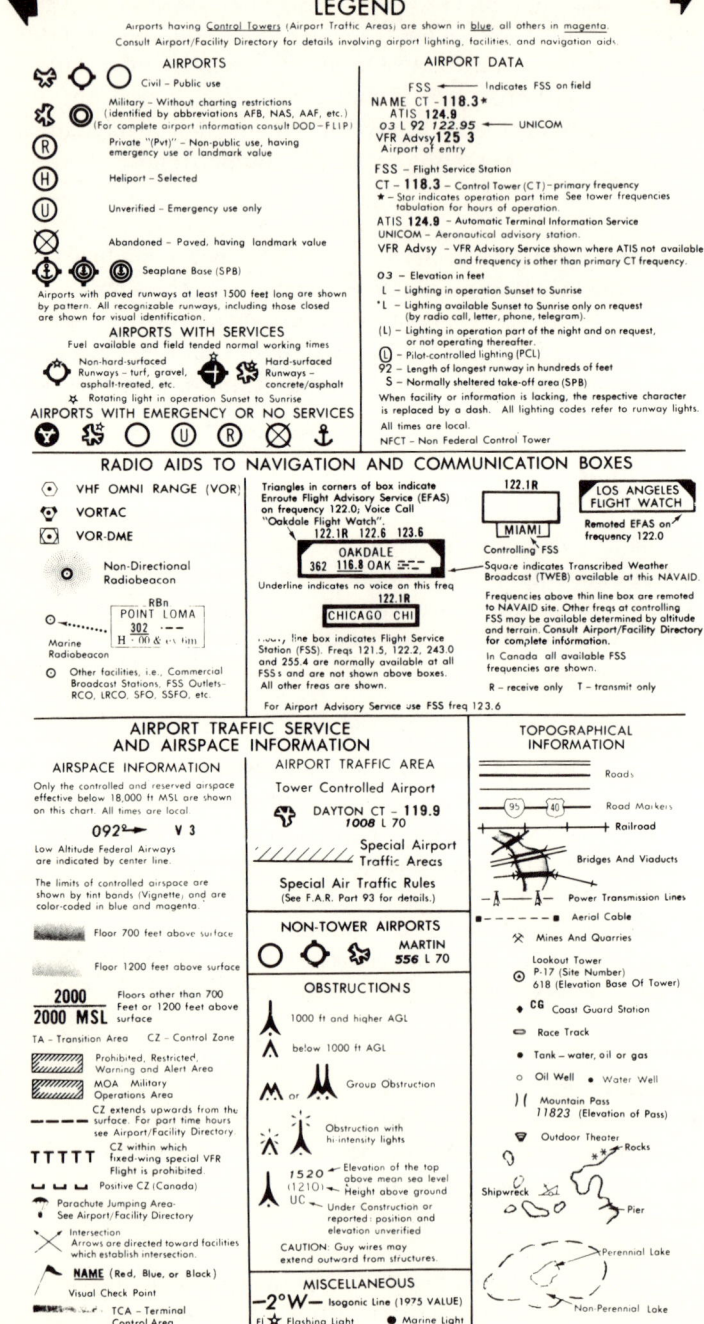

Fig. 16-2. The outer cover of a Sectional Aeronautical Chart with Legend.

Basic Navigational Methods and Procedures

 NEW YORK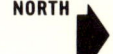
SECTIONAL AERONAUTICAL CHART
SCALE 1:500,000

Lambert Conformal Conic Projection Standard Parallels 41°20' and 46°40'
Topographic date corrected to September 1980

22 ND EDITION *December 25, 1980*
Includes airspace amendments effective *December 25, 1980*
and all other aeronautical data received by November 6, 1980
Consult appropriate NOTAMs and Flight Information
Publications for supplemental data and current information.
This chart will become OBSOLETE FOR USE IN NAVIGATION upon publication of
the next edition scheduled for *JUNE 11, 1981*

PUBLISHED IN ACCORDANCE WITH INTER-AGENCY AIR CARTOGRAPHIC COMMITTEE
SPECIFICATIONS AND AGREEMENTS APPROVED BY
DEPARTMENT OF DEFENSE • FEDERAL AVIATION ADMINISTRATION • DEPARTMENT OF COMMERCE

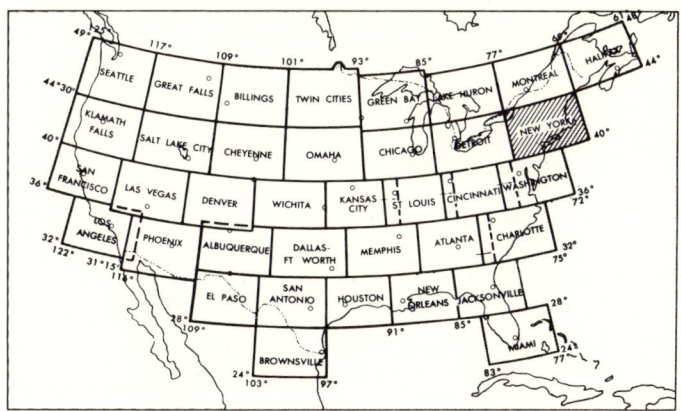

CONTOUR INTERVAL 500 *feet*
Intermediate contour 250 feet
———500——— ———250———

HIGHEST TERRAIN elevation is *6288 feet*
located at 44°16'N – 71°18'W

Critical elevation - - - - - - - - - - - - - - - ▪4254

Approximate elevation - - - - - - - - - - - - ×3200

Doubtful locations are indicated by omission
of the point locator (dot or "x")

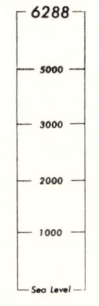

 MILITARY TRAINING ROUTES (MTRs)

All IR and VR MTRs are shown, and may extend from the surface upwards. Only the route centerline, direction of flight along the route and the route designator are depicted – route widths and altitudes are not shown.

Since these routes are subject to change every 56 days, and the charts are reissued every 6 months, you are cautioned and advised to contact the nearest FSS for route dimensions and current status for those routes effecting your flight.

Routes with a change in the alignment of the charted route centerline will be indicated in the Aeronautical Chart Bulletin of the Airport/Facility Directory. Also, the VFR Wall Planning Chart is issued every 56 days and displays current route configurations and a composite tabulation of altitudes along these routes.

Military Pilots refer to Area Planning AP/1B Military Training Route North and South America for current routes.

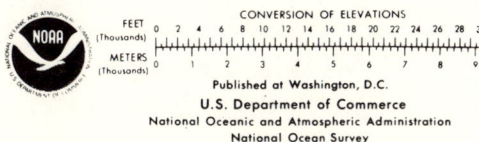

Published at Washington, D.C.
U.S. Department of Commerce
National Oceanic and Atmospheric Administration
National Ocean Survey

Fig. 16-2. The other cover of a Sectional Aeronautical Chart.

Fig. 16-3. The standard cross-country navigational plotter.

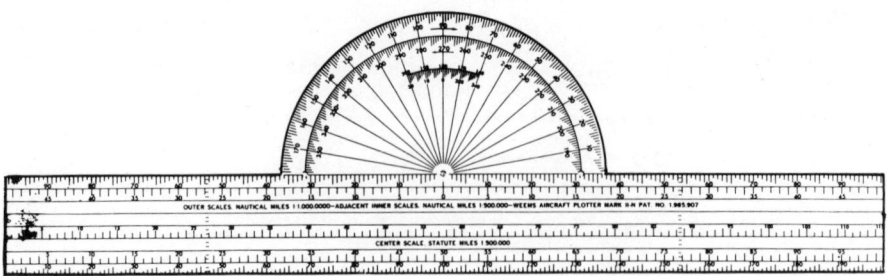

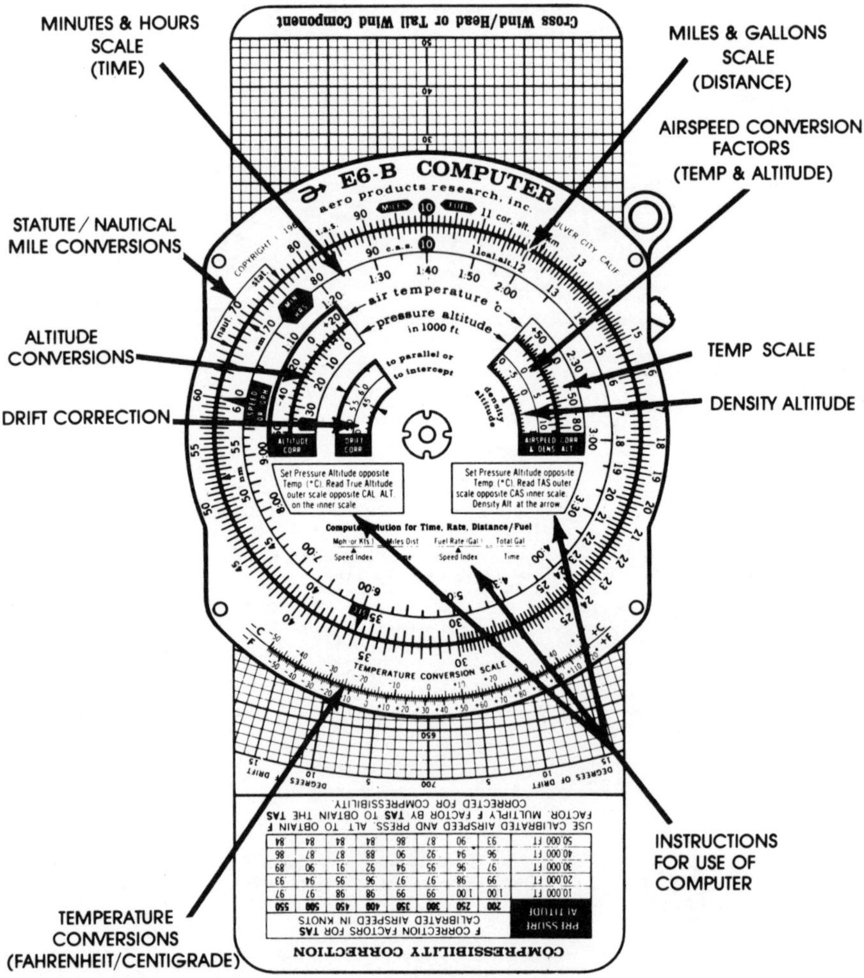

Fig. 16-4. The E6-B type "dead reckoning" navigational computer.

military training areas — because their presence will necessitate another route. See if there are any landing fields along the way. Perhaps you know the area and it has plenty of farmland. Estimate your groundspeed — it'll be equal to the airspeed, only if it's calm. Now you can calculate the time enroute by dividing the air miles between departure and destination by your estimated groundspeed. Knowing your fuel capacity and engine fuel consumption at cruise throttle setting, you'll be able to determine if you have enough fuel to make it non-stop. Always allow a half-hour reserve fuel.

Bracket Your Course

All this means is noting recognizable ground features that lie in a couple of miles to either side of your course. Perhaps it parallels a railroad track, highway or river. These will serve as a crosscheck in keeping you on course. Note prominent features near your destination, in case you fly right over it without seeing it.

Determine True Course

Next, you will want to determine your true course. The Sectional is set up in latitude (horizontal) and longitude (vertical) lines. Lay your plotter parallel to your course and set the center of the protractor portion to intersect a longitudinal line (meridian). The degree line the meridian crosses is your true course.

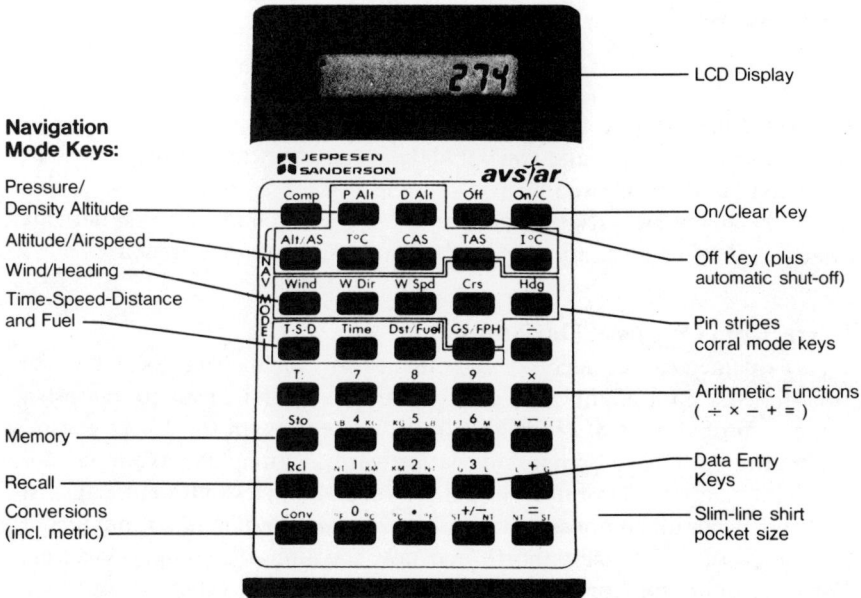

Fig. 16-5. The AvStar electronic navigational computer makes cross-country flight planning easy.

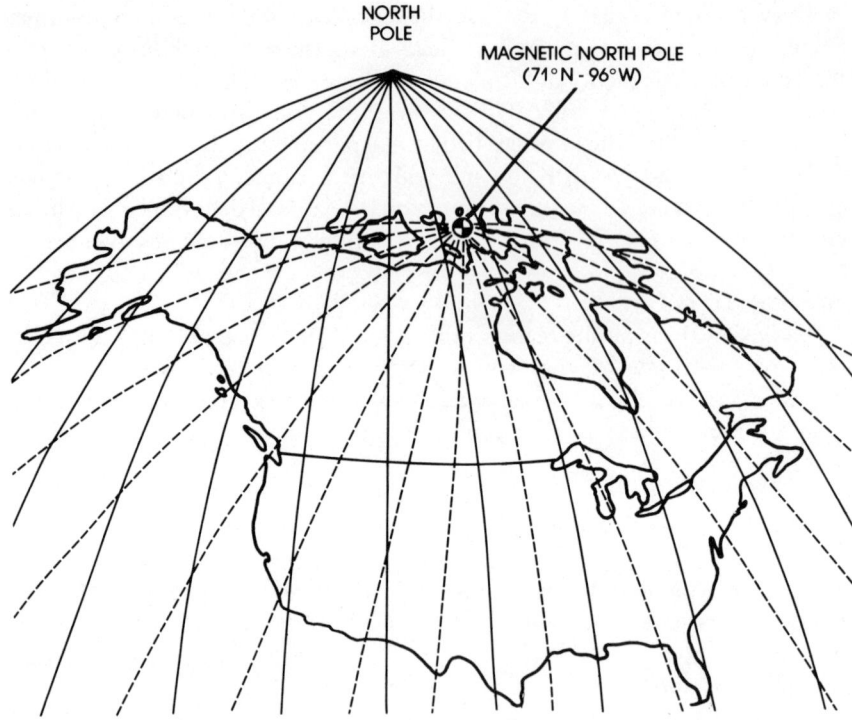

Fig. 16-6. The earth's magnetic pole does not coincide with its geometric pole, giving rise to magnetic variation.

Correct True Course for Magnetic Variation

Check your Sectional and you'll find the Isogonic lines as dashed lines with a degreed number followed by a W or an E. For the Sectional shown in this book, the magnetic variation is near the 10°W Isogonic line. What this means, is that we must add 10° to the true course we determined.

Correct for Compass Deviation

As explained earlier, a compass is affected by iron and steel located in the aircraft. Since ultralights are mostly aluminum and have no radios, a compass probably reads magnetic direction like it ought to. Where general aviation aircraft have compass deviation cards, ultralights have none. An ultralight compass as installed should, however, be checked for error against a known accurate compass. Position the aircraft, with engine running and pilot in place, to magnetic north and note the aircraft's compass reading. Continue doing this for every thirty degrees and you'll generate your own compass deviation chart. These deviations would then have to be applied to the true course and magnetic variation to obtain your magnetic heading.

Basic Navigational Methods and Procedures 137

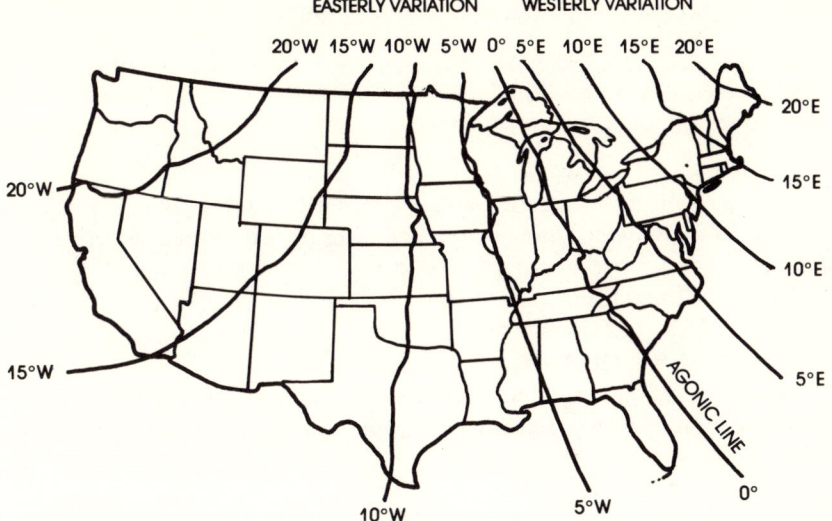

Fig. 16-7. The isogonic lines of the United States show you how to correct compass readings for magnetic variation in your area. The agonic line areas offer no magnetic variation.

Compensate for Wind Drift

Unless you're flying in a calm, or either directly into or with the wind, you'll have to crab to compensate for wind drift. In pilotage, you compensate for wind drift by noting your ground track as compared to your true course. If you're being blown to the left off your course, add some right crab. This will add degrees to your compass heading, so be aware of it. You should be able to determine your approximate drift between your first two checkpoints, and correct accordingly to get back on course and to stay there, noting the new compass heading.

For example, let's assume you have determined your true course to be 90° due east. In south central Pennsylvania, the magnetic course would be 100°. Between your first two checkpoints, you estimate a 10° to the left drift angle. This means you'll have to crab right at 10°, so add 10° to your 100° magnetic course, and you'll arrive at your magnetic heading of 110°. Once back on course, from checkpoint two on, fly with the compass on 110°, and you should reach your destination.

In practice, it is customary to use the following abbreviations:

TC — true course
MC — magnetic course
WCA — wind correction angle
TH — true heading
VAR - magnetic variation
MH — magnetic heading
DEV — compass deviation

We can summarize the above example as follows:
TC — 90°
VAR — 10°W
MC — 100°
WCA — 10°R
MH — 110°

As long as you are flying cross-country, you'll need to continually solve the above problem of determining course and heading. The standard method for converting true course to compass heading is as follows:

TC + WCA = TH
TH + VAR = MH
MH + DEV = CH

Always remember to add westerly variation and subtract easterly variation. An easy way to remember is by saying to yourself: "east is least and west is best"!

Groundspeed, Elapsed Time and Fuel Burned

Only by knowing the performance capabilities of your ultralight, can you determine these figures. Let's suppose you are flying the ultimate ultralight, with the following specifications and performance figures:

Cruise Speed — 5.8 mph indicated
Fuel Capacity — 5 gallons
Fuel Consumption — 2 gallons per hour
Endurance — 2½ hours, including reserve
Rate of Climb — 750 feet per minute

First of all, you must determine your true airspeed. Remember, the air gets thinner as you climb, making your ASI read low, dropping 2% for each 1,000

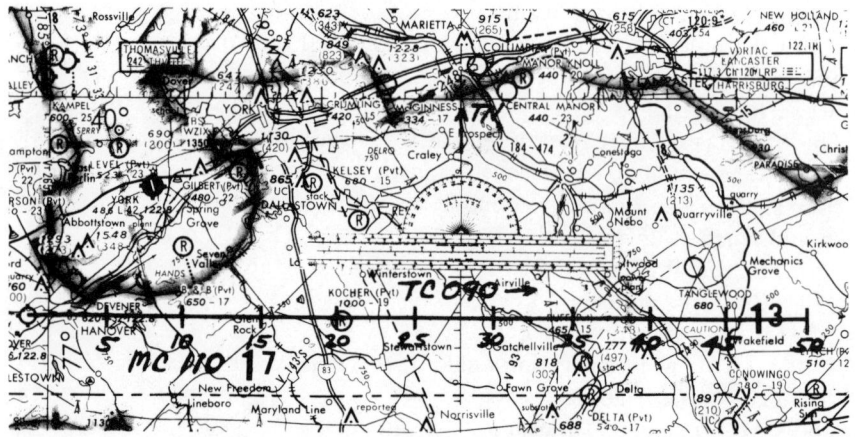

Fig. 16-8. True course plot of a 50 mile cross-country flight on the New York Sectional, as described in the text.

Basic Navigational Methods and Procedures

feet. Checking your course plot, you determine that a cruising altitude of 1,500 feet AGL (above ground level) will be safe. Since the ground is already 500 feet MSL (above main sea level), your cruising altitude is 2,000 feet MSL. Therefore, you must add 4% to your indicated airspeed to arrive at your TAS. The answer is approximately 60 mph. That means each 5 mile checkpoint distance would take 5 minutes, while the entire flight would take about 50 minutes, in a calm.

But, you sleep-in and, by the time you get to the field, the wind is blowing at 15 mph, 45° from the right of your course, i.e., at 135°. It's going to require crabbing and your flight will take a little longer. Taking off at 9 AM, you climb to 2,000 MSL (1,500 AGL). It takes 2 2/3 minutes to reach altitude, and a total of 7½ minutes to reach your first checkpoint, 5 miles out. Your average groundspeed was only 40 mph because of the climbout and headwind component.

Looking at the Sectional and checkpoint two, you notice you are drifting to the left, and estimate it to be about 10°. You immediately start turning to right until you intersect your intended course, at which point you crab 10° right and maintain a magnetic heading of 110°. And, since your compass has no deviation, your compass heading is also now 110°.

By the time you get to checkpoint two, everything looks fine. You're on course, and it looks like you should pass over checkpoint three, judging by the landmarks in front of you, to either side of course. At this point, you also notice your elapsed time from checkpoint two to three took about six minutes, and you figure your groundspeed to be 50 mph. That 15 mph wind, 45° to the right has given you a 10 mph headwind component.

Knowing your actual groundspeed, you can now calculate how long it'll take you to fly the rest of the way. Having gone by three checkpoints, that's 15 miles covered, leaving 35 miles to go. Dividing 35 miles by 50 mph tells you it'll take almost three-quarters of an hour to finish the flight — 42 minutes, to be exact. The entire trip out then, will take 61.5 minutes (7½ climb to first checkpoint, and 54 minutes to fly from checkpoint one to the destination), instead of the original 50 minutes. Since you carry 2½ hours of fuel, you'll have a reserve of 88 minutes.

If the wind conditions remain the same for the return trip, you'd have a tailwind component of 10 mph, making the total trip time approximately 1¾ hours. You won't have to fill-up before returning, as you'll have 45 minutes of reserve when you return to your home field.

Dead Reckoning

The term is derived from the phrase reckoning by deduction, and it's really an extension of pilotage. Its use depends on the three primary flight instruments ASI, altimeter and compass, as well as on known wind conditions. It can be used to solve the three basic navigational problems of knowing your position, how to get to a new position, and when you'll reach that position.

Dead reckoning allows you to know what heading to fly before you takeoff. By using the principle of the wind triangle, any basic ultralight navigational problem can be solved.

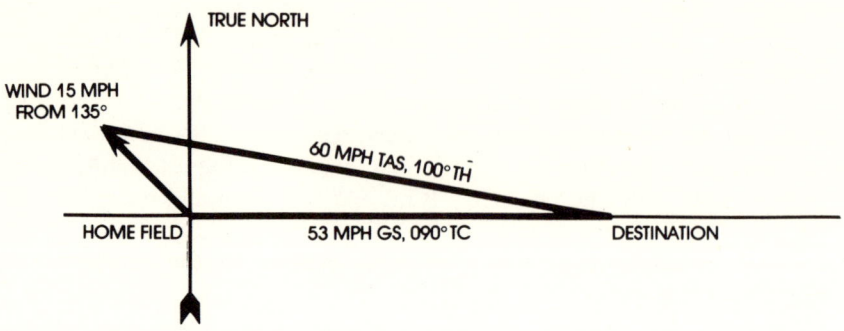

Fig. 16-9. Wind triangle for a flight to a point 50 miles due east, as plotted on the Sectional in Fig. 17-8.

Constructing a Wind Triangle

In order to get a clear understanding of what's involved here, let's use the same set of performance figures and flight conditions as we did in the pilotage example, with a few pertinent additions:

Cruise Speed — 58 mph (60 mph TAS)
Fuel Capacity — 5 gallons
Fuel Consumption — 2 gallons an hour
Endurance — 2½ hours
Rate of Climb — 500 feet per minute
Cruising Altitude — 2,000 MSL
Wind at 2,000 MSL — 10 mph. @ 135°
TC — 90°

Given the above information, we can find groundspeed, WCA, compass heading, time enroute, ETA and fuel required.

Starting with a clean sheet of paper, show true north and south by drawing a vertical line. Halfway down the line, put a point — that will be your departure field. Lay your plotter on the line with its straight edge at your point and draw a line to represent your TC of 90°. From the same departure point, draw another line from 135° to represent the wind direction.

Now, using the statuate mile scale on the bottom of your plotter, measure up the wind line (vector) 15 mph and mark the point. Place your plotter's statuate mile scale "0" at the end of the wind vector and swing it until the 60 (your TAS) intersects the TC line. Draw a line connecting the end of the wind vector with the TC line. Measure the length of the line from your original origin to the TC-TAS intersection — that's your groundspeed and it reads 53 mph! You can also measure your TH as 100°, with the protractor part of your plotter.

Basic Navigational Methods and Procedures

Going back to our original data, we can determine our compass heading:
TC + WCA = TH = 100°
TH + VAR = MH — 100° + 10°W = 110°
MH + DEV = CH = 110° = 0 = 110°
CH = 110°

Since our groundspeed is 53 mph and our destination is 50 miles away, we can calculate our time enroute as 50 miles divided by 53 mph equals .94 hours, or 57 minutes. Adding 2.67 minutes for climbout, gives a total flight time of about 60 minutes. Having 2½ hours of fuel on board, we can safely fly the distance.

As a point of interest, go back to the pilotage example, and you'll see that our figures are in real close agreement. But, the point is, the dead reckoning approach is more accurate, gives you more assurance after you're in the air, and allows you to enjoy the flight more. It let's you do your figuring on the ground, and have more fun flying.

The Flight Back

To determine the particulars of your return flight, all you have to do is construct another wind triangle, as before. This time, your ground speed is up to 69.75 mph, since the crosswind now has a tailwind component. The compass heading is determined as follows:

TC + WCA = TH - 260°
TH + VAR = MH - 260° + 10°W = 270°
MH + DEV = CH = 270° + 0
CH = 270°

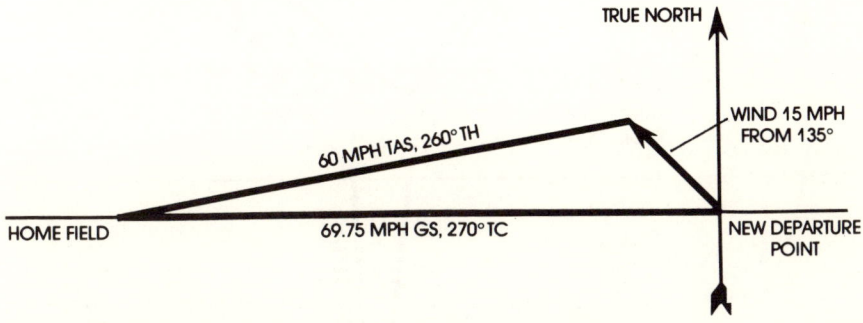

Fig. 16-10. Wind triangle for the return flight, as plotted on the Sectional in Fig. 17-8.

Navigating a Triangular Course — The Airspeed Circle

Navigation of a triangular course could be done by constructing three separate wind triangles, but there's an easier way — let's call it the airspeed circle method. For our example, we'll plot a course from Reigle to Farmer's Pride to Bendigo to Reigle, as shown on the portion of the Sectional in this book, Fig. 2-43.

First draw a true north-south line in the center of your paper and then a true east-west line thru the center of the first line. From that intersection, draw in your wind speed and direction (15 mph and 135°). Now, get a circle compass, and spread it to a distance equal to your TAS. Put the compass point on the end of the wind vector and draw the circle.

Now, let's analyze each leg of the triangle.

Leg One, Reigle to Farmer's Pride, TC measures 34°. Draw in the 34° line from the intersection of your true N-S and true E-W lines until it hits the airspeed circle. The length of that TC line will be your groundspeed for Leg One, which equals 41 mph. Now draw a line from the point where TC 1 intersects the airspeed circle, back to the tip of the wind vector — that's your TH 1, and it measures 48°. That means your CH 1 would be 58°.

Now analyze Legs Two and Three the same way, and there you have it. A summary of the data from the airspeed circle is as follows:

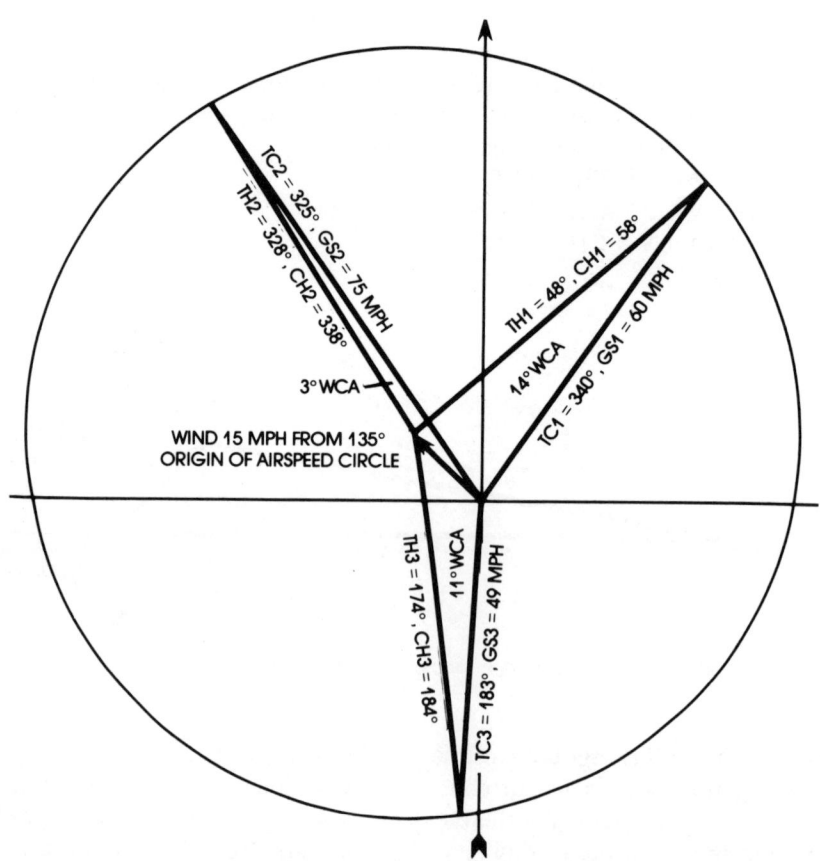

Fig. 16-11. Airspeed circle with wind triangles for the three leg flight plotted on the Sectional in Fig. 17-1.

Basic Navigational Methods and Procedures 143

Item	Leg One	Leg Two	Leg Three
Distance	12.5	10°	18.5
TC	134°	325°	183°
WCA	14°	3°	11°
TH	48°	328°	174°
CH	58°	338°	184°
GS	60 mph.	75 mph.	49 mph.
Time Enroute	13 min.	8 min.	23 min.

The total enroute time is 44 minutes. Adding 6 minutes for climb to and descent from cruising altitude, gives a total flying time of 50 minutes. There won't be any problem in making the flight non-stop.

As an aid to navigational computations, there are various flight computers available. The two basic kinds are the E6B sliderule type and the electronic type. They all come with complete instructions, and make navigation almost fun, if not easy.

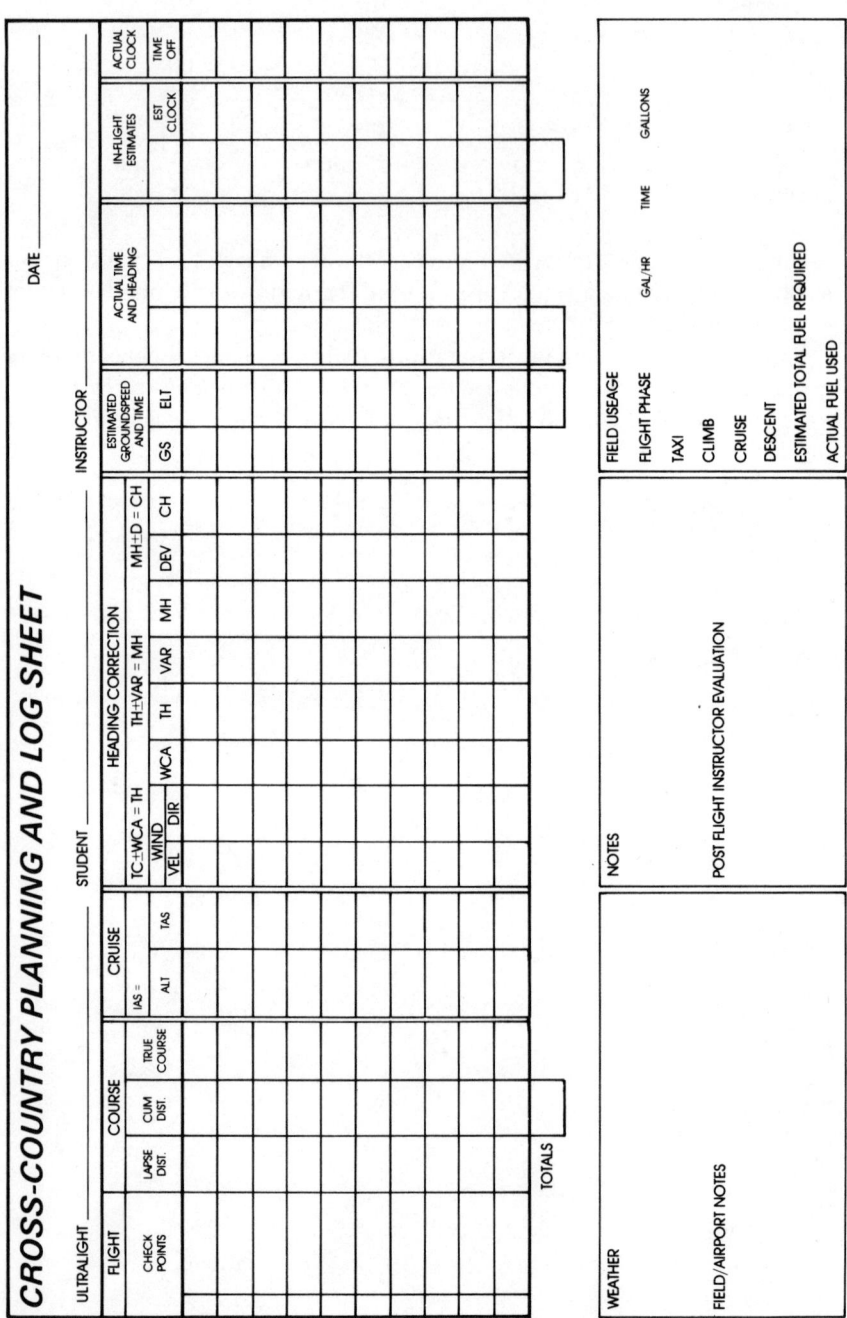

Fig. 16-12. The Cross-Country Planning and Log Sheet is essential for flights away from home field.

17
Wind Chill Factors and Hypothermia

The average normal temperature at which the human body is designed to function is 98.6°F; to maintain that temperature, the body produces heat from metabolic processes and may absorb heat from the environment, if available. Hypothermia results when the body is unable to generate enough heat and the internal temperature drops below normal.

The body can lose heat through evaporation, radiation, conduction and convection. Awareness of body heat loss is important to ultralight pilots.

Evaporation of perspiration is the natural means by which the body cools itself. As water changes from liquid to gas, much heat is dissipated.

Radiation is another means of losing body heat; it occurs when uncovered skin is exposed to the atmosphere. During winter, however, our clothing usually protects us from this type of heat loss. The only potential trouble spot is an uncovered head. For those who do not wear a hat, this is a major cause of heat loss.

Conduction is the transfer of heat from the body to another surface. Touching very cold metal, for example, can draw heat from the body. Generally this is not a major means of heat loss for ultralight pilots.

Of primary concern to ultralight pilots is convection, the transference of body heat to the air. If the rate of air moving across the body increases, convective heat loss increases at a disproportionate rate. This supercooling is called wind chill.

Ultralight pilots are subject to wind chill because they sit totally exposed to the wind. There are three types of wind that should concern ultralight pilots: surface wind; the relative wind encountered in flight; and the air blast or slipstream caused by propellers.

Once your ultralight is in the air, you no longer have to worry about the cooling effects of surface wind. Instead, relative wind becomes the concern. Relative wind is the air that strikes an aircraft in flight.

Unlike the winds caused by atmospheric phenomena, the speed of this relative wind is the same as the airspeed of the ultralight; so a pilot flying at 40 mph is in effect sitting in a 40-mph wind. Without an enclosure, the ultralight pilot has no way to escape the effects of relative wind except by landing.

AIRSPEED OR WIND IN MPH	OUTSIDE AIR TEMPERATURE (°F)											
	50	40	30	20	10	0	-10	-20	-30	-40	-50	-60
	EQUIVALENT TEMPERATURE (°F)											
calm	50	40	30	20	10	0	-10	-20	-30	-40	-50	-60
5	48	37	27	16	6	-5	-15	-26	-36	-47	-57	-68
10	40	28	16	4	-9	-21	-33	-46	-58	-70	-83	-95
15	36	22	0	-5	-18	-36	-45	-56	-72	-85	-99	-112
20	32	18	4	-10	-25	-39	-53	-67	-82	-96	-110	-124
25	30	16	0	-15	-29	-44	-59	-74	-88	-104	-118	-133
30	28	13	-2	-18	-33	-48	-63	-79	-94	-109	-125	-140
35	27	11	-4	-20	-35	-49	-67	-82	-98	-113	-129	-145
40	26	10	-6	-21	-37	-53	-69	-85	-100	-116	-132	-148

(wind speeds greater than 40 mph have little additional effect)	LITTLE DANGER (for properly clothed person)	INCREASING DANGER	GREAT DANGER

Danger from freezing of exposed flesh.

Fig. 17-1. The wind chill factor must be known when the ambient air temperature drops to 50°F. Open-air ultralight flight could be extremely dangerous for an inadequately clothed pilot, especially when climbing several thousand feet.

The Wind Chill Factor Chart

Wind chill was investigated by the U.S. Navy in the Antarctic in the winter of 1941. The resulting wind-chill chart illustrates the perceived drop in temperature due to wind chill with actual thermometer readings. (See Fig. 17-1).

Down the side of the chart are estimated wind speeds in knots and miles per hour; ultralight pilots should consider these wind speeds to be airspeeds. Across the top of the chart are actual thermometer readings in Fahrenheit. These correspond to ambient or outside air temperature. For most ultralight

Wind Chill Factors and Hypothermia

pilots, the temperature at ground level is adequate for calculating wind chill. High-flying winter pilots should be aware of the standard lapse rate, the 3.5°F drop in temperature for every 1,000 feet of altitude, and deduct this from surface temperature.

The center of the chart gives the equivalent temperature, or wind chill factor, caused by wind. For example, if the outside air temperature were 20°F and a pilot were to fly at 40 mph, there would be a drop of 40°F. This means that, from the pilot's seat, the temperature would feel like -21°F.

Across the bottom of the chart is an estimate of the danger of these temperatures to "properly clothed" persons. Little danger means little danger of exposed flesh freezing—frostbite. Medium danger means that flesh may freeze within one minute of exposure. Great danger means that flesh may freeze within 30 seconds of exposure. These frostbite conditions *must* be avoided.

Pilots of ultralights that are designed so that the pilot sits on the propeller blast or is close enough to a pusher propeller to have a significant amount of air flow past them, must consider more than just flying speed. They should be aware of the supercooling effect of the additional "wind" caused by the propeller. To estimate this effect, the ultralight should be tied down tightly and an airspeed indicator placed where the pilot's head would be while in flight. To estimate wind chill, the airspeed reading at cruise rpm should be added to the airspeed at which the pilot plans to fly. As noted on the chart, winds of more than 40 mph have little additional effect.

The Five Stages of Hypothermia

Hypothermia can be prevented with physical activity. However, ultralight pilots, sitting still in their aircraft, have little chance for vigorous activity. Therefore, we must respect the cold and wear proper clothing.

There are five stages of hypothermia, each characterized by different symptoms as body temperature drops. When the body begins to lose heat faster than it can produce it, blood circulation is altered to preserve heat in the internal organs. Kidneys, liver, heart and brain must be as near a normal temperature as possible to maintain their functions. Blood flow at or near the surface of the skin is reduced, which cuts the rate of heat transfer to the atmosphere.

The first stage, when body temperature drops from 98.6°F to as low as 96°F, is evidenced by intense shivering. Shivering is the body's attempt to produce heat by muscular contraction. During the second stage, when body temperature ranges from 95°F to 91°F, the victim suffers from violent shivering, thinks and speaks sluggishly and sometimes suffers amnesia. He loses judgment and reasoning power but does not realize this is happening. Motor response decreases, and the pilot may lose grip of the controls.

The most important sign of the third stage, when body temperature drops to between 90°F and 86°F, is that shivering stops. Skin is blue and puffy, coordination is poor, and thinking is muddled. Muscles are rigid; there may be some erratic jerking. The victim of the third stage of hypothermia is only fuzzily aware of his environment.

The fourth stage, when body temperature drops to between 86°F and 78°F, is evidenced by even poorer reflexes followed by unconsciousness. The fifth stage, when body temperature drops below 78°F, most often results in cardiac arrest and death.

Treating Hypothermia

Treatment of hypothermia depends first on recognizing its symptoms. Never ignore shivering. As your body drains its energy reserves to produce heat, body temperature will drop steadily. The treatment is to reduce the amount of exposure and add heat to the body.

If you start to feel very cold, land as soon as possible. Since your head is the primary source of heat loss, keep it covered at all times. Exercise mildly to get warm, but not enough to perspire. Remember, perspiration is the best way to lose heat. If you exercise enough to become exhausted and then stop, even for a short time, your body's rate of heat production will drop instantly to half or less. Violent incapacitating shivering may begin. If your skin is already damp with sweat, you may be just minutes away from unconsciousness.

Rubbing the extremities of someone suffering from an advanced stage of hypothermia is not a good idea. This brings relatively warm blood to the surface where it is cooled by the colder tissues. The cooled blood then returns to the core of the body and compounds hypothermia.

If you can, go someplace warm and protected, preferably a house, and add heat to your body. Warm drinks help a lot, as do energy-producing foods such as candy bars. If your clothes are wet, get out of them and into dry ones or into a dry sleeping bag.

If the victim is wet and semi-conscious, help him strip and into a sleeping bag. You should strip also and get into the bag with him. This skin-to-skin contact is the most effective means of treating an advanced case of hypothermia. While it may sound sexy to some, just beware that the victim could be nauseated and throw up.

If hypothermia has developed to the third stage, where shivering stops, medical help must be called immediately.

Frostbite is a possibility with advanced hypothermia. Superficial frostbite occurs when the outermost layers of skin freeze—often fingers, toes, nose, lips and ears. The areas feel cold and tingle with pain. They look waxy and white but are still soft when pressure is applied. Treat superficial frostbite immediately by warming the affected part inside your clothing, armpit or groin. Do not massage it or rub it with snow.

Deep frostbite occurs when the deeper tissues also freeze. There is a loss of feeling and absence of pain, and you may not notice the onset of deep frostbite. The areas will be firm or hard under pressure. This tissue will die, and blisters will form and later the tissue may sluff off.

The most damage will result if the tissue is partially thawed and allowed to refreeze. Do not try to thaw it by touching hot objects or keeping it near a fire. If the part cannot be thawed properly, leave it frozen until you can obtain proper care. It is already frozen and, if carefully protected, there will be no additional damage. Leave blisters unopened, cover the parts with a clean soft dressing, and seek medical help immediately.

Deep frostbite can be thawed by immersing the area in water of 110°F or slightly cooler. The tissues should not be rubbed or massaged. Analgesics may be needed for pain. By the way, do not smoke, as that can cause vascular spasms. Again, seek medical help immediately; hospitalization may be necessary in some cases.

Most cases of hypothermia develop in ambient temperatures of 30°F to 50°F. That is without adding the wind chill factor, so hypothermia is not just a winter problem. It can occur during the cool calm mornings that are so good for ultralight flying, during cool evenings, or at higher altitudes. But as long as you are aware of the potential, clothe yourself properly, can recognize the symptoms and act immediately to correct the situation, the threat of hypothermia does not have to interfere with the safety of your flying adventures.

"An aircraft is considered to be under control when the pilot is able to direct its course at will. This is the case as long as the aircraft remains unstalled, and is moving in the desired direction at the selected altitude."

18
Ultralight Equipment — Glossary and Function

Even before the Wright Brothers first flew at Kitty Hawk, terms were being developed to describe the various parts of an aircraft and its operations. We will explain the more important terms to make the text more meaningful.

aerobatic FAR (Federal Aviation Regulations) 91.71 defines aerobatic flight as an intentional maneuver involving an abrupt change in an aircraft's altitude, an abnormal attitude or abnormal acceleration not necessary for normal flight. Aerobatics are characterized by maneuvers such as loops, rolls, spins and inverted flight. An aircraft must be able to withstand ultimate load factors of +9.0 and -4.5 to be considered aerobatic. Ultralights are generally not designed for aerobatic flight.

aerodynamic center The point on an airfoil where the pitching moment coefficient is constant from zero lift up to near the stall. It is usually located

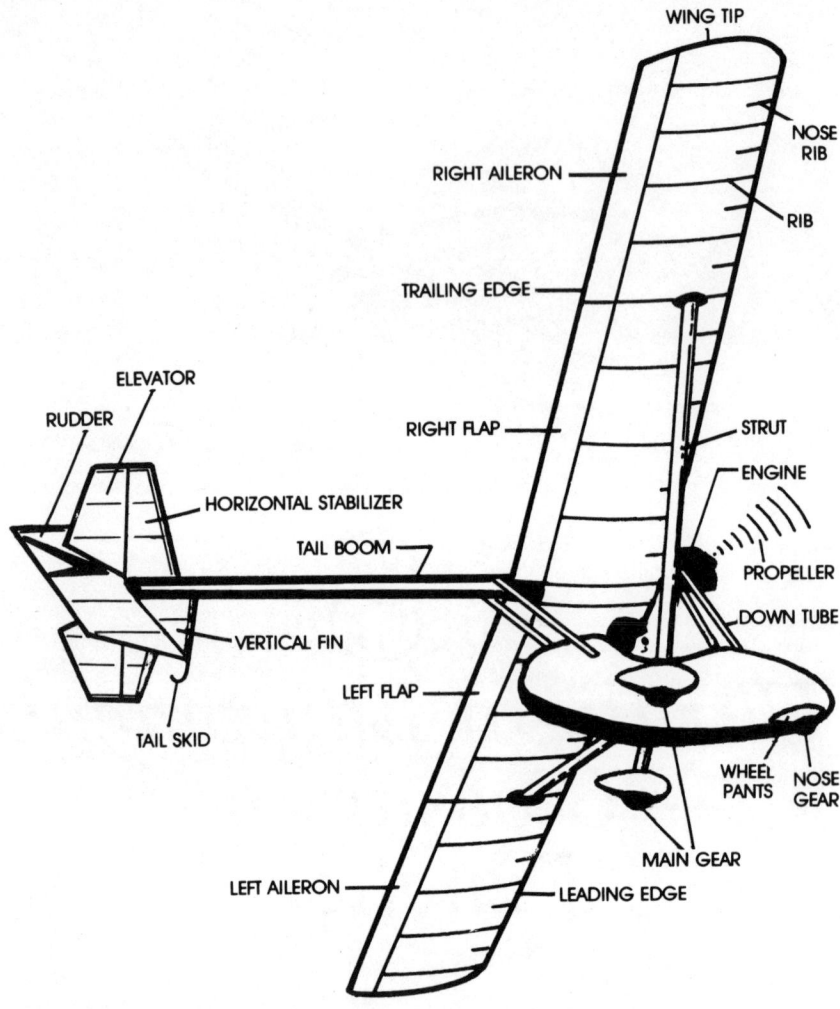

Fig. 18-1. The primary parts of an ultralight air vehicle.

one quarter of the chord length behind the leading edge, and the lift and drag are assumed to act through it.

aerodynamic coefficient   A non-dimensional number which is the ratio of an aerodynamic force or moment, divided by the dynamic pressure and the wing area (and chord for moments); principally the coefficients of lift, drag, and pitching moment.

aerodynamic damping   The aerodynamic forces that resist rotations about an aircraft's axes, and those that reduce rotations after a gust or control input is made.

aerodynamic force   The force generated by a body's motion through the air.

AGL   Above ground level.

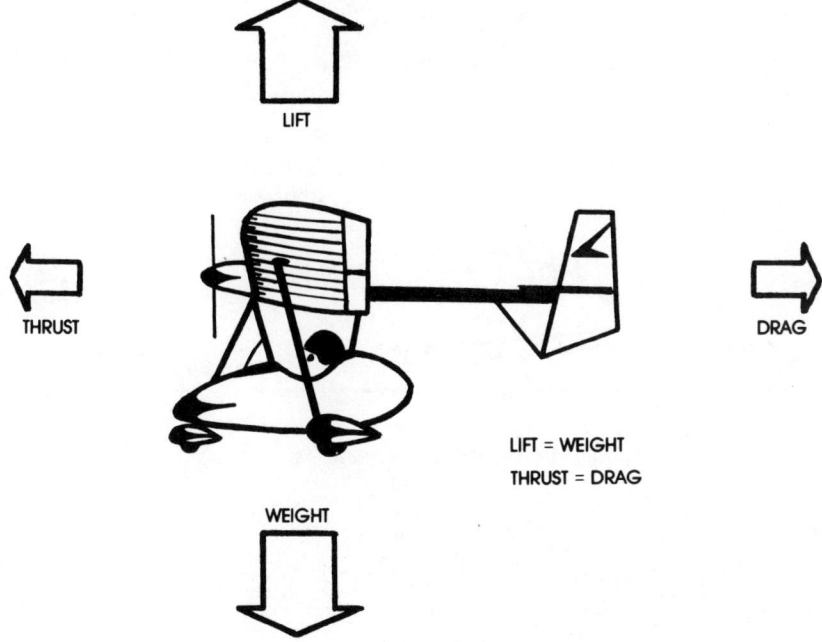

Fig. 18-2. The four main forces acting on an ultralight in straight and level flight.

aileron Plain flap control surfaces located at a wing's outer trailing edge. They deflect asymmetrically to rotate an aircraft about its longitudinal axis.
aileon, free Similar to a stabilator type elevon.
air A mechanical, not chemical, mixture of gases that comprise the earth's atmosphere. The main components are oxygen (78%) and nitrogen (21%) the rest being made up of several inert gases. The exact content of the mixture varies with altitude and latitude.
aircraft Any and all structures that are supported in the air either by its dynamic reaction with the air or by a buoyant gas. More specifically, the tern applies to free and tethered balloons, blimps, dirigibles, kites, ornithopters, flying model airplanes, hang gliders, ultralights and other fixed wing aircraft, helicopters and autogyros.
airflow The movement of air. It is measured as a flow of mass per unit time.
airfoil Any structure designed to gain a useful reaction from the airflow around it. More specifically, the cross sectional shape of a wing, tail surface, or propeller.
airframe The structural and aerodynamic components that support the loads transmitted to it by the parts in contact with the air. They include: fuselage, wings, tails, landing gear, tail booms, fairings, nacelles, and control surfaces.
airplane Any engine powered, heavier-than-air, fixed wing aircraft that generates lift solely by the dynamic reaction of the air with its surfaces.

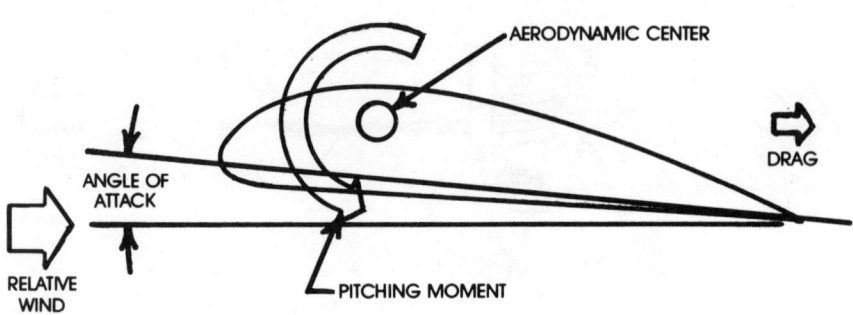

Fig. 18-3. The forces acting on a lifting airfoil.

airscoop The open end of a duct, hood, cowl or nacelle that projects into the airflow to carry a portion of that airflow to another part of the aircraft, such as an engine cooling air intake.
airspace The air above a certain area on the earth, typically identified by some sort of boundary.
airspeed The speed with which an aircraft moves relative to the air.
airstream The relative airflow near a body.
airworthy When an aircraft is suitable for safe flight.
altitude The vertical distance of an aircraft above the surface. Altimeters are usually set with sea level being zero, and all navigational charts are zeroed to sea level.
amphibian An aircraft capable of operating off land and water.
angle of attack The angle a wing chord line makes with respect to the relative wind.
angle of climb The angle between the local horizon and climb path of an aircraft.
angle of incidence The angle the wing chord line is set with respect to an airplane's longitudinal axis.
aspect ratio The slenderness of a wing defined as wingspan squared divided by wing area. Wingspan divided by average chord also yields aspect ratio.
atmosphere The gaseous envelope surrounding the earth, through which flight occurs. It extends up to about 700 miles above the surface, thinning out with altitude. 50% of the atmosphere is below 18,000 feet.
attitude The orientation of an aircraft's axes with respect to the horizon.
aviation The art and science of manned flight, especially with heavier-than-air vehicles.

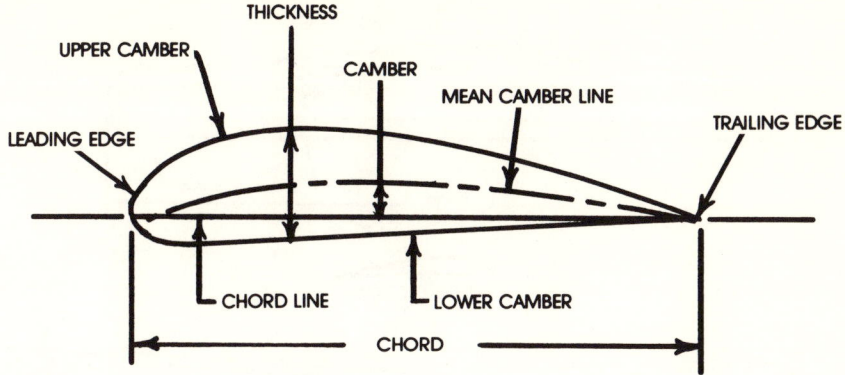

Fig. 18-4. Nomenclature of the typical airfoil.

axes Three imaginary straight lines that pass perpendicularly through an aircraft's center of gravity, around which the aircraft rotates. The axes include: vertical (yaw axis), longitudinal (roll axis), and lateral (pitch axis).
balloon The momentary, inadvertent rise of an airplane as it tries to touch down for a landing. It is normally caused by the pilot over-controlling the landing, and can be amplified by ground effect.
bank The tilt of an aircraft about its longitudinal axis, which is necessary to turn.
Bernoulli's Principle This most fundamentally important aspect of low speed aerodynamics was discovered in 1738 by a Swiss physicist named Daniel Bernoulli. He found that in any given airflow, the sum of the static and dynamic pressures was always the same, everywhere in the flow. What this means is that for any streamline flow wherever the velocity is high the pressure is lowered and, wherever the velocity is lowered the pressure is raised. For example, the air striking the leading edge of a wing is stopped at a stagnation point, resulting in a higher pressure. Also, whenever an airflow meets with a restriction, as in a venturi, the velocity must increase to move the same volume of air through it.
biplane An airplane with two sets of wings, one set above the other.
boundary layer The thin layer of air immediately adjacent to a surface over which air flows, composed of decelerated velocity air—the slowest air being nearest the surface. The boundary layer can be laminar—where the velocity deceleration is gradual, or turbulent—where the various velocities mix with one another.
burble The airflow condition above and behind a stalled wing. Turbulence.
camber The depth of an airfoil's mean line, expressed as a percentage of the chord. The curvature of an airfoil.
canard Aircraft with a small lifting wing located ahead of the main wing. The small lifting wing itself, is also referred to as a canard.
ceiling, absolute The highest altitude attainable by an aircraft.

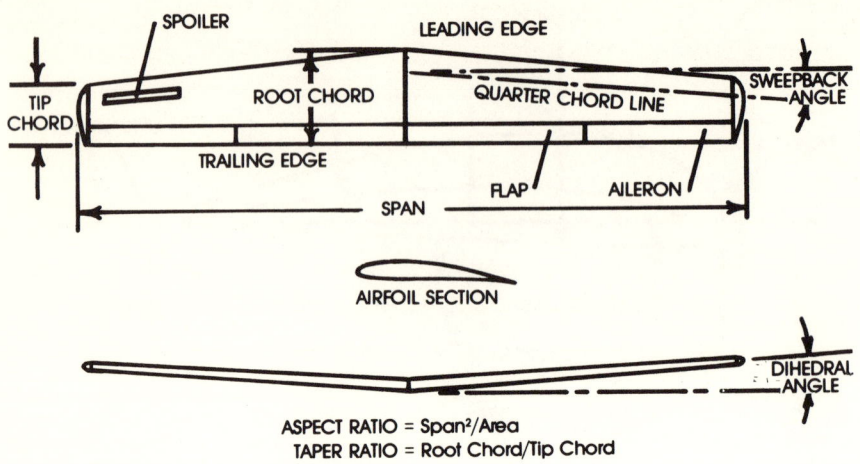

Fig. 18-5. The defining geometry of a wing.

ceiling, service The altitude where the maximum rate of climb is 100 feet per minute.
center of gravity (CG) The point where an aircraft balances and, through which the force of gravity is considered to act. It is also the point where the three axes meet perpendicular to one another and, around which all motions are considered to occur. A proper CG is most important to stable flight. Too rearward a CG will result in instability and could lead to an unrecoverable spin. Too forward a CG could make a landing flare impossible.
center of pressure The point on a surface where the resultant of all the aerodynamic forces can be considered to act.
centripetal force The acceleration on a turning aircraft that causes it to turn. Specifically, it is the inward tilt of the wing's lift vector which counteracts the centrifugal force tending to prevent the aircraft from turning.
chord The length of an airfoil or wing section, as measured from leading edge to trailing edge.
circulation The motion of air rotating about an axis. Specifically, the air rotating about a lifting airfoil as a theory of lift.
Coanda effect The tendency of an airstream to attach itself to and follow the surface over which it flows. The surface should be smooth and rounded.
compression strut Structural member joining leading and trailing edges.
control surface Aerodynamic surfaces that move to control the altitude, speed and direction of an aircraft. These include: elevator, ailerons, rudders, flaps, spoilers, trim tabs and drag brakes.
crab Technique used to maintain course in a crosswind. It is established by a turn done in the normal way with aileron, while rudder is used to correct any adverse yaw. The aircraft will appear to be travelling sideways with respect to the ground but, it will be going straight with respect to the wind. Controls are then neutralized.

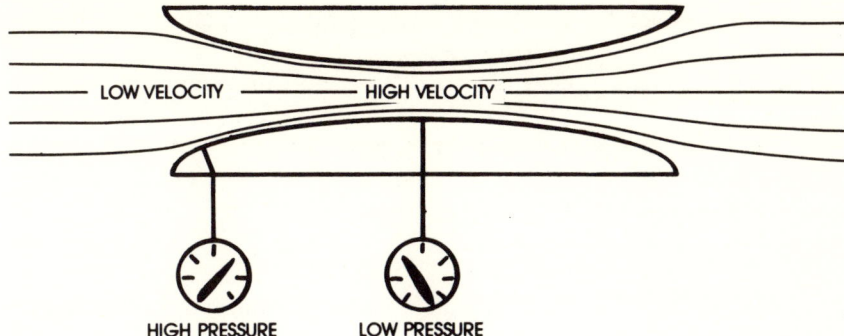

Fig. 18-6. Air flowing through a venturi is forced to speed-up, while the pressure shows a corresponding decrease.

cruise speed The speed, below top speed, at which an aircraft normally flies for reasons of fuel economy and engine life. It generally occurs at throttle settings of 65% to 75% of wide open throttle.
delta wing Triangularly shaped wing with swept back leading edges and a straight trailing edge.
dihedral Small angle made between the horizontal and the wing panel chord plane.
downwash The downward flow of air behind a lifting wing. It is caused by the wing airfoil and angle of attack, and results in an aircraft being lifted.
drag The total aerodynamic force tending to retard the motion of an aircraft in a direction parallel to the airflow. It develops whenever an object moves relative to the air.
dynamic pressure The pressure developed when air comes to rest at the front of an object. It is measured in pounds per square foot and is often referred to as "q," which is the product of one-half the air density times the velocity squared.
dynamic stability The oscillating motion an aircraft exhibits in returning to its original flight trim condition.
eddy A small rotational turbulence within an airflow.
elevator The aft hinged portion of the horizontal stabilizer, or canard, which controls pitch attitude and airspeed.
elevon Dual purpose control surface on flying wings, used for both pitch and roll control.
empennage The horizontal and vertical tail group.
empty weight The weight of an aircraft unloaded and unoccupied, including: usable fuel, undrainable oil, engine coolant, hydraulic fluid and attached ballast weights.
engine Power source for flight, typically a reciprocating two cycle.
engine cowling A fairing positioned around the engine, designed to direct cooling air through the engine in an aerodynamically efficient manner.

equilibrium A condition of balance. An aircraft is in equilibrium when lift equals weight and thrust equals drag.
fairing A streamlined shell placed over a structure or component to reduce drag.
fin The fixed leading portion of the vertical tail.
fineness ratio A body's length divided by its thickness. Optimal fineness ratios develop minimal drag for a given thickness.
fixed-pitch propeller A propeller with blades that are unadjustable in flight, as is typical for ultralights.
fixed-wing aircraft An airplane with a stationary wing and not a helicopter or rotary winged aircraft.
flare A maneuver done just before making contact with the ground in landing. The nose is brought up to slow the aircraft while minimizing the sink rate until touchdown.
flap The hinged, moveable inboard trailing edge portion of a wing used to alter camber, and therefore lift and drag, especially during landing.
flight Movement of an aircraft through the atmosphere.
flutter A harmonic oscillation or vibration of an aircraft part or control surface caused by aerodynamic forces acting on a too flexible or improperly balanced component.
fluid Any liquid or gas.
flying wing An aircraft that houses all essentials for flight within the wing. It does not have a horizontal tail.
flying wires Cables located beneath the wing that support it against positive flight loads.
free stream The airflow outside the region affected by the passage of an aircraft.
fuselage The part of an airframe that houses the pilot, with wings, tail and landing gear attached to it.
g The acceleration subjected to an airplane by gravity and abrupt maneuvers. One g is generated in straight and level flight, due to the acceleration of gravity, which is 32 feet per second, acting on the mass of the aircraft.
g-load Forces developed due to the acceleration of an aircraft, as compared to the forces on it under straight and level, unaccelerated flight in calm air. They are developed by gusts, turns and abrupt maneuvers.
gas The fluids that expand indefinitely, having no definite shape or volume.
glide Flight with little or no thrust, characterized by a loss in altitude.
glider An engineless aircraft that derives its thrust from gravity.
glide ratio Numerically equal to the lift over drag (L/D) ratio, it is the horizontal distance an aircraft will travel (engine-off) for every foot of altitude lost.
gross weight An aircraft's total flying weight, including: empty weight, fuel, oil, pilot, passengers and cargo.

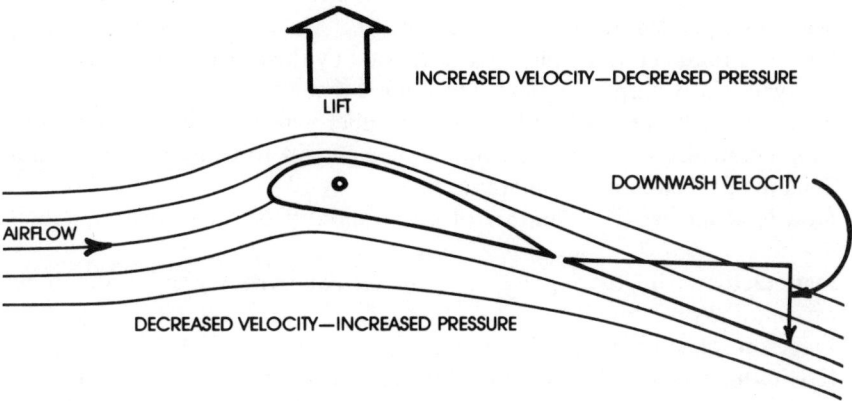

Fig. 18-7. Lift is the reaction of an airfoil to deflecting the air that flows over it in a downward direction.

ground effect The tendency for an aircraft to "float" when within a half wingspan of the runway. It's caused by a reduction in induced drag, and less power is required to fly this close to the ground.
hang glider Engineless aircraft from which the pilot is suspended in a harness or swing seat. It is typically controlled by weight shifting, but may incorporate aerodynamic controls as well. They are normally foot-launched and landed.
heading The direction an aircraft is pointing toward. Commonly the magnetic compass heading.
induced drag That portion of the drag due to the generation of lift. It is associated with wing tip vortices and is greatest at low speeds.
joystick The control stick. It is connected to ailerons (and/or spoilers) and elevator on three axis controlled aircraft and, to rudder and elevator on two control aircraft.
kilo One kilogram, which equals about 2.2 pounds.
kinetic energy The energy possessed by a body due to its motion.
kite Generally, an aircraft flown while tethered to the ground or a ground based vehicle. It flys solely by the relative wind pressure against its surfaces. Also, colloquilly, a Rogallo-type hang glider.
laminar flow Airflow characterized by streamlines that move smoothly over one another, without mixing.
landing gear The undercarriage on which an aircraft rests while on the ground. Typically wheels, it can be floats and skis, as well.
landing speed The airspeed with which an aircraft touches the ground, typically equal to the stall speed.
landing wires Cables located above the wing that support it against landing loads and negative flight loads.
lateral axis The imaginary line passing from wingtip to wingtip, through the center of gravity. Pitch motions occur about it.
lateral stability Stability about the roll axis.

leading edge The front of a wing.
lift The total aerodynamic force generated by the dynamic reaction of the air against a wing perpendicular to the relative airflow.
lift coefficient The non-dimensional number which represents the lift of a wing or airfoil. It is obtained by dividing the lift force by the free stream dynamic pressure and wing area.
limit load factor The number of g's an aircraft is designed to handle in flight.
load factor The total g load on an airplane from gravity and other accelerations.
longeron The primary longitudinal structural member of a fuselage.
longitudinal axis The imaginary line passing from nose to tail through the center of gravity. Rolling motions occur about it.
longitudinal stability Stability about the pitch axis.
maneuvering speed The highest allowable speed for abrupt maneuvers or very rough air. It is typically set at twice the stall speed and is designed to protect against structural failure. Here, the aircraft could receive a maximum 4 g load, where the wing would stall before loading the airframe further.
mean aerodynamic chord (MAC) The average chord of the wing, or wings, which can be used to represent an aircraft's center of aerodynamic forces.
mean line The center line of an airfoil section equidistant between the upper and lower surfaces.
minimum flying speed The lowest airspeed attainable, out of ground effect.
moment A force times its distance from the center of action (rotation).
momentum The quantity of motion, expressed as mass times velocity.
monocoque A type of fuselage construction in which the skin carries all or most of the stresses.
mush A nose high condition of flight between minimum power required speed and minimum level speed. It is flight on the "back side of the power curve."
navigation Act of directing flight from point to point.
negative g A condition that loads an aircraft from the top.
oscillation Vibration or movement to either side of a neutral point.
parasite drag Drag caused by components that do not contribute to lift. It becomes more important with speed, increasing as the square of the airspeed.
parasol wing A monoplane with its wing mounted above the fuselage.
pitch A rotation about the lateral axis, as in nose-up, nose-down.
pitching moment The moment about an aircraft's lateral axis. A nose-up pitching moment is positive.
phugoid A long period longitudinal oscillation where an aircraft flies along a "roller-coaster" path with little or no change in angle of attack.
planform The top view of an object, particularly a wing.
power loading Pounds gross weight divided by horsepower.
potential energy The energy possessed by an object due to its height.

Ultralight Equipment - Glossary and Function

pressure altitude Height in the atmosphere as measured from the standard pressure datum of sea level.
profile The cross section of a body, especially the airfoil section of a wing.
propeller Airfoil shaped rotating wing used to convert torque into thrust. Larger diameters and lower tip speeds are more efficient.
propeller pitch The distance a prop moves forward for each revolution.
propeller wash The airflow generated behind a rotating propeller.
range Maximum distance an aircraft can fly at a given cruising speed, with a 45 minute reserve.
rate of climb Vertical speed of an aircraft. Typically quoted as a maximum at standard sea level conditions.
relative wind In the study of aerodynamics, it makes no difference whether an aircraft is flying through still air, or air is being blown past a stationary aircraft. The aerodynamic forces generated are identical.
Reynolds number The basic law of airflow similarity which is the ratio of dynamic forces divided by viscous forces. Airflow patterns are the same for similar Reynolds numbers.
rib The part of a wing that provides the airfoil shape.
Rogallo wing A flexible membrane sail-type wing as used on most hang gliders.
rotate The pitch rotation required during the takeoff run, enabling an aircraft to lift off.
rudder Moveable part of vertical tail used to yaw aircraft. It balances adverse aileron yaw in independent three-axis control aircraft. Helps steer the aircraft while taxiing. It is NOT analogous to a ship's rudder and is not used like one!
ruddervator Dual purpose surface on V-tailed aircraft used to control pitch and yaw.
sailplane High performance glider with a glide ratio of 25 or better.
scale effect The affect of size on aerodynamic reactions. Generally, the larger an aerodynamic body, the more efficient it can be.
seaplane Aircraft designed for operations off water only.
separation When air fails to follow the contour of the object over which it is flowing. It occurs behind bluff bodies and more important, on the upper surface of wings at high angles of attack.
sink rate The vertical speed of descent, typically quoted as the minimum rate for the aircraft.
slat High lift device, located at the leading edge, that directs high velocity air over a wing to delay the stall.
slot The gap between a slat and a wing.
slope lift Wind striking vertical terrain generates this vertical wind on which gliders soar.
span Wing length from tip to tip.
spar A wing's primary load bearing member.
spin An aerobatic maneuver or a condition where an aircraft is stalled and

rotating in a small radius about a vertical axis. The nose is pointed well below the horizon, as well.

spoiler Control surface located on wing's upper surface, which normally lies flush. When deflected, it reduces lift and increases drag. They roll the aircraft when deflected singly, and control glide path when deflected simultaneously.

square-cube law As a body is increased in size, its mass grows as the cube of its increase, while its area grows only as the square. In other words, doubling the linear dimension of an airplane will result in eight times the weight, and only four times the wing area, provided the construction and materials are similar. This also implies a doubling of the wing loading.

stabilator Control surface, similar in function to elevons, but mounted below and behind a wing's trailing edge, as on the Mitchell flying wings. Also, an all-moving horizontal tail.

stagger On a biplane, the relative longitudinal spacing between leading edges. Positive stagger is when the top wing is forward, while negative stagger is when the bottom wing is forward.

stagnation point The forward location on a body where the airflow comes to rest, as on the leading edge of a wing section.

stall A breakdown in the lift of a wing which occurs when the airflow separates from the upper surface at an angle of attack peculiar to the airfoil section, Reynolds number and aspect ratio. It can happen at any airspeed, depending on the angle of bank and g-loading. For normal, straight and level, unaccelerated flight it occurs at a given airspeed, which can be no higher than 27 mph, according to FAA rules.

static pressure The ambient pressure of the atmosphere, normally 14.7 psi at sea level.

streamline The path traced by air molecules as they move over an object.

strut Commonly a tube used to externally brace a wing to a fuselage.

sweepback The angle of the wing leading edge (or quarter chord line) makes with the lateral axis.

taildragger An aircraft with main wheels in front and a tail wheel, or skid.

tail skid Rear fuselage member used to protect the tail from the ground.

tail wheel Rear wheel on a taildragger.

thermal A parcel of heated air that breaks away from the ground and rises, usually generated over dark areas of the earth. Gliders use the updraft to gain altitude.

thrust The propelling force generated by a propeller, needed to move an aircraft and equal its drag force. Under full throttle, its greatest value occurs at zero airspeed (i.e., static conditions).

thrust line The imaginary centerline of the thrust, which is typically near or at the aircraft's center line.

tip dragger Wing tip control on some ultralights used to yaw the aircraft when actuated singly, and to control glide path when actuated simultaneously.

tip stall The stalling of a wing tip. It could occur at higher lift coefficients

and develop into a spin. A properly designed wing should not tip stall—the stall should begin at the center of the wing.
track Path of an aircraft over the ground.
trailing edge Aftermost portion of a wing.
tricycle landing gear Aircraft landing gear with a nose wheel and two main wheels aft the center of gravity. This is the most popular type, and it is stable.
turbulence The mixing of streamlines in an irregular "eddying" motion, such as is found behind bodies that have experienced a flow separation. Also, uneven movement of the atmosphere found mostly near the surface and behind obstructions. It can be strong enough to destroy an aircraft.
ultralight An aircraft with an empty weight under 254 pounds, a stall speed of 27 mph, and a top level speed of 63 mph, according to FAA definition.
vector A force or velocity that has both magnitude and direction. They can be added graphically by joining heads to tails or mathematically by trigonometry.
velocity never exceed (Vne) The red-line speed of an aircraft, beyond which structural damage may occur in calm air.
venturi A tube with a smaller diameter in the center than at the ends. When air, or any fluid, moves through a venturi, the velocity increases in the center, while the pressure decreases.
vertical axis Imaginary line which passes vertically through an aircraft's center of gravity, and about which the aircraft yaws.
viscosity The "stickiness" of air as evidenced by its tendency to adhere to the surfaces over which it flows, as well as to itself.
vortices Organized circular flow of air caused by pressure differences in an airflow. Examples include the "bound vortex" equivalent of a wing in circulation theory, and vortices developed at a wing tip.
washout Upward set of wing tip trailing edge to minimize tip stall.
weight and balance The practice of keeping an aircraft's gross weight and center of gravity within prescribed limits as established by the designer. Deviation from this results in poor performance and possibly catastrophic instability.
wind milling A propeller being turned by the airstream and not the engine.
wind tunnel Basically, a tube with a fan installed in one end to draw air over an object mounted inside the "test section." The object is connected to a sensitive balance system in order to measure aerodynamic forces and moments.
wing loading Gross weight divided by wing area. The greater its value, the higher the stall speed. Doubling wing loading increases the stall speed by 41%!
wing tip vortices Horizontal "tornados" emanating from the wing tips of all aircraft, especially at high angles of attack. The heavier the aircraft, the stronger the vortices. They develop because the high pressure air on the bottom of a wing tends to flow around the tips to the low pressure air on top of a wing.

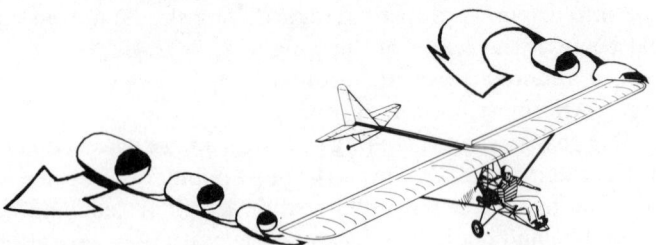

Fig. 18-8. The air flowing from the tips of a lifting wing rolls out and down in a series of vortices that increase with angle of attack.

Appendices

"Pulling back on the stick will not stretch your glide during an approach."

The Ultralight Solution — Separating Aircraft from Air Vehicles

Item: *On April 11, 1981, the captain of a B-727 reported a "small, tan-colored object with wings" passing within 300 feet of his airliner as he prepared to land at Sky Harbor Airport in Phoenix, AZ.*
Item: *On July 17, 1982, the pilot of a single-engine aircraft on an IFR flight between Cincinnati and Charleston, WV reported a near collision with an unidentified flying object that passed within 100 feet of his airplane at 7,000 feet in IFR conditions. Evasive action was taken to maintain separation.*

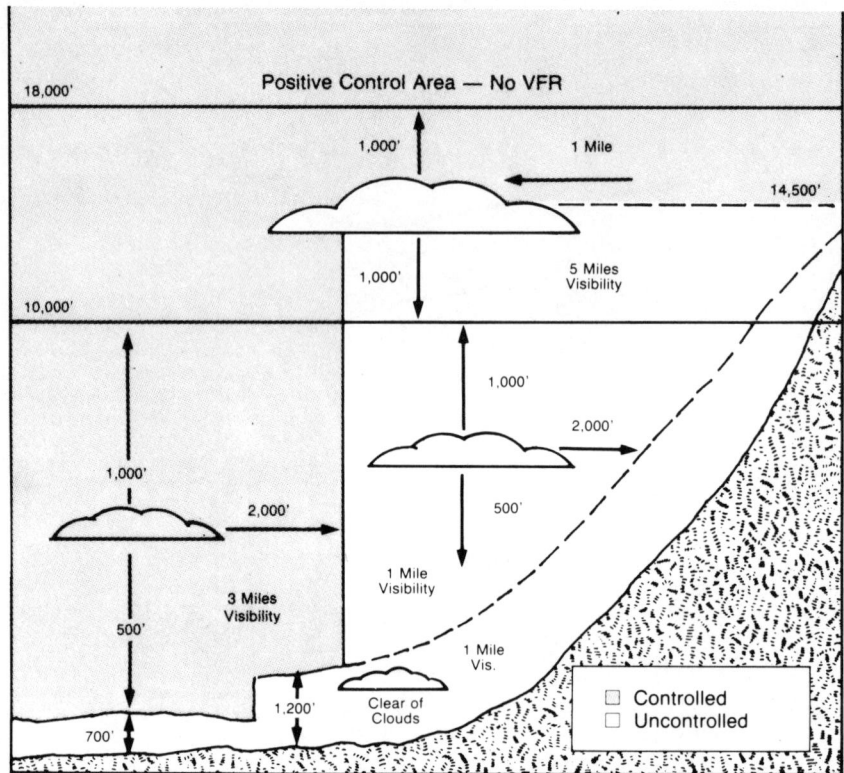

Fig. A-1. Visibility and cloud clearance minimums.

In both instances the flying objects were identified not as a bird, a kite or a UFO, but as a variety of "hang glider," a type of flight vehicle whose antecedants are almost as old as civilized man. Human beings have suspended themselves from winged contraptions to glide down from walls, towers, and mountain tops for hundreds of years, without causing any particular disturbance. Then on December 17, 1903, the Wright Brothers successfully launched their "Flyer" from ground level with the aid of a gasoline engine, and the aircraft was born. This new form of transportation temporarily relegated the foot-launched glider to the realm of antiquity, where it remained until the 1970s, when it was reborn as a sport.

Thanks to new ultra light materials derived from space age technology, plus the availability of lightweight chainsaw or snowmobile gas engines, the modern hang glider has evolved into a miniature airplane, some small enough to be folded up and transported on a man's back or on top of a car, others efficient enough to attain altitudes of over 16,000 feet. Given the low initial cost, when constructed from a kit, and the miserly fuel consumption, this is now where it is at—especially for the young at heart. The appeal of surfing, or sky diving, or even motorcycling, has paled before the excitement

The Ultralight Solution - Separating Aircraft from Air Vehicles

of zipping around the sky in a magical vehicle.

Until recently FAA was content to consider all hang gliders, even those with engines, as "non-aircraft" as long as they were capable of being footlaunched. But their sudden increase in popularity and their appearance in busy airspace created a situation that has had to be faced. An estimated 30,000 are now in use in this country alone, with manufacturers delivery projections for the decade of the 1980's running into the hundreds of thousands. Random activities in busy airspace by uncontrolled flight vehicles which have no radio and show up only weakly at best on radar could result in tragic accidents.

The FAA solution has been to create a new category of flying machine — the *ultralight vehicle*. According to newly enacted Federal Aviation Regulation Part 103, an ultralight is any flying vehicle which:
1) is designed for single occupancy only
2) is intended for sport or recreational use only
3) does not have an airworthiness certificate
4) if unpowered, weighs less than 155 lbs.
5) if powered, weighs less than 254 lbs. and
   - has a top speed of no more than 55 knots (63.25 mph) in level flight
   - has a power-off stall speed no greater-than 24 knots (27.6 mph)
   - has a fuel capacity not exceeding five U.S. gallons

Flying vehicles which meet these definitions are to remain free of any federal regulations as regards design, construction, maintenance, airworthiness, registration, or pilot certification. This hands-off policy reflects the agency's willingness to have private individuals "self-regulate" in their own interests, under the leadership of industry and cooperative associations.

Under the new rule, the safety of the airspace will be maintained by limiting ultralight operations in terms of airspace access, right of way, flights over persons on the ground, hours of operation, and weather conditions.

Persons who operate a vehicle under Part 103 will be responsible for ascertaining that they are in compliance with all aspects of the rule. This will also require an understanding of FAA's airspace structure, as described in various aeronautical publications. In addition, ultralight operators as well as aircraft pilots should become familiar with practices established at uncontrolled airports for the separation of their vehicles during takeoff and landing. Because of their extreme sensitivity to turbulence, ultralights are highly vulnerable to wingtip vortices (swirling currents of air) from even the smallest aircraft.

At this time FAA does not anticipate providing guidance on flying techniques, but will offer helpful information on airspace procedural matters and on establishing flying sites. Operators are also referred to the publications of their manufacturer or to the ultralight divisions of the Aircraft Owners and Pilots Association (AOPA), and the Experimental Aircraft Association (EAA); to the Powered Ultralight Manufacturers Association (PUMA), and to the U.S. Hang-Gliding Assoc.

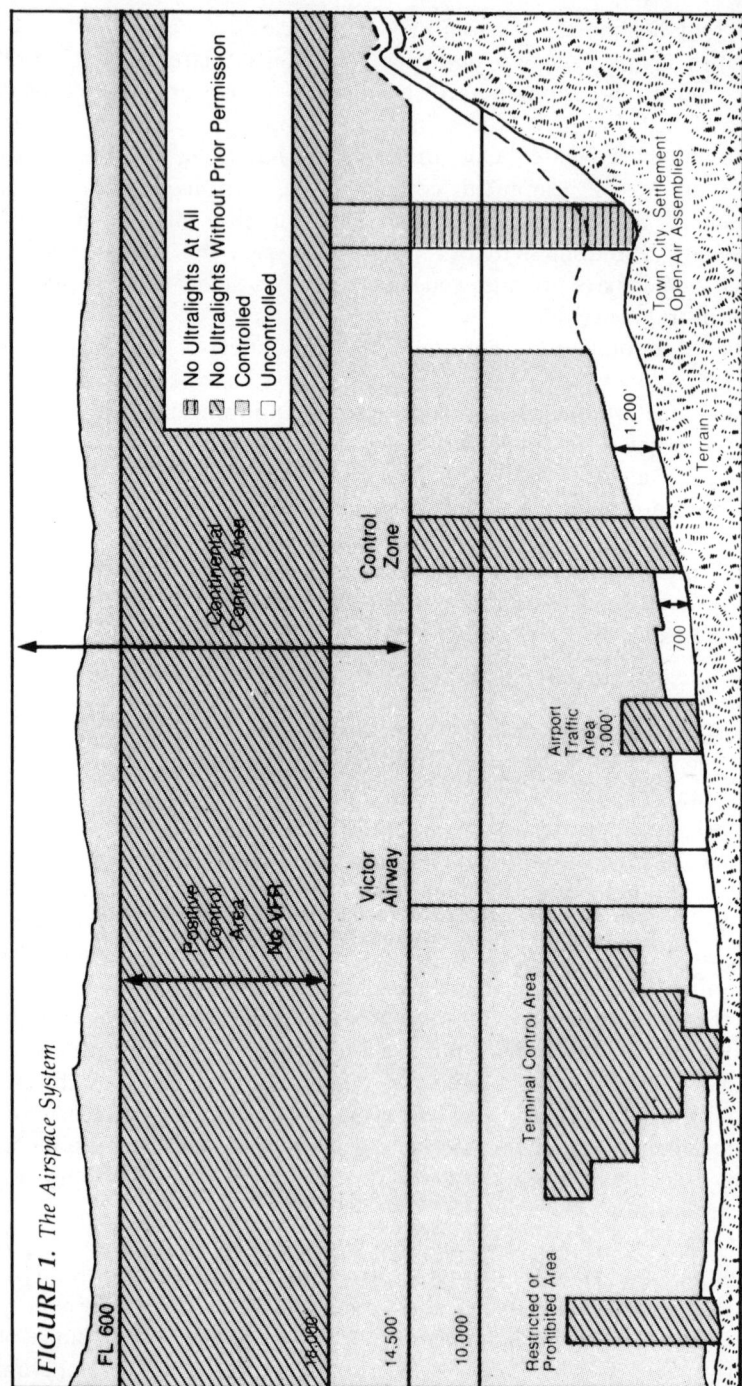

Fig. A-2. Maintaining VFR conditions for ultralight pilots requires minimum distances from clouds in different kinds of airspace, as well as minimum visibility and constant reference to the surface.

A-2
The Ultralight Rules — FAR Part 103

Title 14 — Aeronautics and Space
CHAPTER I — FEDERAL AVIATION ADMINISTRATION
DEPARTMENT OF TRANSPORTATION
[Docket No. 21631; New Part 103]
PART 103 — ULTRALIGHT VEHICLES
Operating Requirements

INTRODUCTION
AGENCY: Federal Aviation Administration (FAA), DOT.
ACTION: Final rule.
SUMMARY: This amendment establishes rules governing the operation of ultralight vehicles in the United States. The rule defines ultralight vehicles in two categories; powered and unpowered. To be considered an ultralight vehicle, a hang glider must weigh less than 155 pounds, while a powered vehicle must weigh less than 254 pounds; is limited to 5 U.S. gallons of fuel;

must have a maximum speed of not more than 55 knots; and must have a power-off stall speed or not more than 24 knots. Both powered and unpowered ultralight vehicles are limited to a single occupant.

Those vehicles which exceed the above criteria will be considered aircraft for purposes of airworthiness certification and registration, and their operators will be subject to the same certification requirements as are aircraft operators. These rules for ultralight vehicles are needed to achieve an acceptable level of air safety by reducing potential conflict with other airspace users and to provide protection to persons and property on the ground.

The rule governs the operation of ultralight vehicles by specifying the airspace which requires prior authorization of Air Traffic Control (ATC), prohibiting operations over congested areas, and providing for operations during twilight hours with proper lighting. Right-of-way and minimum visibility rules are also established.

The FAA has chosen not to promulgate Federal regulations regarding pilot certification, vehicle certification and vehicle registration, preferring that the ultralight community assume the initiative for the development of these important safety programs. The ultralight community is expected to take positive action to develop these programs in a timely manner and gain FAA approval for their implementation. Should this approach fail to meet FAA safety objectives, further regulatory action will be necessary.

EFFECTIVE DATE: October 4, 1982.

FOR FURTHER INFORMATION CONTACT: Ken Peppard, Airspace and Air Traffic Rules Branch (AAT-220), Federal Aviation Administration, Washington, DC 20591, (202) 426-3128; Gary Perkins, General Aviation Operations Branch (AFO-820), Federal Aviation Administration, Washington, DC 20591, (202) 426-8194.

BACKGROUND

The FAA issued Advisory Circular No. 60-10, entitled "Recommended Safety Parameters for Operation of Hang Gliders" on May 16, 1974. That advisory circular contained recommended safety parameters for the operation of sport hang gliders, in lieu of formal Federal regulation. The advisory circular defined "hang glider" as "an unpowered, single-place vehicle whose launch and landing capability depends on the legs of the occupant and whose ability to remain in flight is generated by natural air currents only."

The sport of hang gliding has advanced dramatically since Advisory Circular No. 60-10 was issued. There is now widespread use of powerplants, landing gear and movable control surfaces to increase the speed, altitude and distance capabilities of the vehicles. Many models have passenger-carrying capability. As a result of those developments, many hang gliding vehicles no longer fall within the scope envisioned by Advisory Circular No. 60-10. The addition of powerplants and controllable aerodynamic surfaces has created vehicles which can approximate the operational capabilities of fixed-wing and rotary-wing aircraft.

The increasing performance capabilities of these vehicles and their greatly increased number, have created a potential hazard to other aircraft and operators, as well as to the ultralight operators themselves. As the result of aerodynamic improvements, many unpowered hang gliders are now capable of extended soaring to altitudes exceeding 10,000 feet above the point of launch and distances of over 100 miles. The powered hang gliders now have the capability of sustained flight above 10,000 feet and forward speed exceeding 50 knots.

The operations of these vehicles are now a significant factor in aviation safety. The vehicles are routinely operated, without authorization, into regulated airspace, such as airport traffic areas, terminal control areas, positive control areas, and prohibited and restricted areas. Many operations have also taken place over congested areas and spectators and into adverse weather conditions in which operations may be conducted by pilots and aircraft which are qualified for instrument flight (IFR conditions). The midair collision potential presented by unauthorized operations is contrary to the FAA responsibility of ensuring the safety of all airspace operations including air carrier aircraft.

To illustrate the potential for hazardous situations that can arise, the FAA has recorded data detailing numerous instances of ultralight vehicles in controlled airspace causing near-miss situations with aircraft. The following examples highlight the problem:

(1) On March 24, 1982, an MU-2 flew between two ultralights operating off the end of the runway at Winter Haven, Florida. Both ultralights were equipped with floats and were operating at night without lights.

(2) On April 11, 1982, a Western Airlines 727 captain reported a near-miss with an ultralight vehicle in the vicinity of Phoenix Sky Harbor Airport.

(3) In May of 1981, the pilot of a single-engine aircraft reported a near miss with an ultralight vehicle near Paso Robles, California. According to the report filed under the FAA Aviation Safety Reporting Program, the ultralight was operating at 7000 feet in IFR weather conditions. The airplane pilot, who was operating on an IFR flight plan, was forced to take evasive action to avoid a collision.

To establish regulations to deter flights which present a serious danger to aircraft and to provide a basis for necessary enforcement action, the FAA published Notice of Proposed Rulemaking No. 81-6 on July 27, 1981 (46 FR 38472). That notice proposed to include both powered and unpowered hang gliders under the generic term "ultralight vehicle" and included proposed weight and fuel limitations for those vehicles. The notice proposed a number of operational limitations for ultralight vehicles, while recognizing that the vehicles are used primarily for sport purposes.

More than 2500 persons and organizations submitted comments to that proposed rule. This rule is the result of FAA consideration of those comments in light of its responsibility for safety in the National Airspace System. Because of the growing significance of this segment of the aviation community, the new rules have been codified under a new part of the Federal Aviation Regulations, Part 103.

REGULATIONS
ADOPTION OF THE AMENDMENT

Accordingly, the Federal Aviation Regulations (14 CFR Chapter I) are amended, effective October 4, 1982 by adding to Subchapter F (14 CFR Chapter I) a new Part 103 as follows:

Subpart A — General

Sec.
103.1 Applicability.
103.3 Inspection requirements.
103.5 Waivers.
103.7 Certification and registration.

Subpart B — Operating Rules

Sec.
103.9 Hazardous operations.
103.11 Daylight operations.
103.13 Operation near aircraft; right-of-way rules.
103.15 Operations over congested areas.
103.17 Operations in certain airspace.
103.19 Operations in prohibited or restricted areas.
103.21 Visual reference to the surface.
103.23 Flight visibility and cloud clearance requirements.

PART 103 —
ULTRALIGHT VEHICLES

Subpart A — General

§ **103.1 Applicability.**

This part prescribes rules governing the operation of ultralight vehicles in the U.S. For the purposes of this part, an ultralight vehicle is a vehicle that:

(a) is used or intended to be used for manned operation in the air by a single occupant

(b) is used or intended to be used for recreation or sport purposes only

(c) does not have any U.S. or foreign airworthiness certificate, and

(d) if unpowered, weighs less than 155 pounds; or

(e) if powered:

(1) weighs less than 254 pounds empty weight, excluding floats and safety devices which are intended for deployment in a potentially catastrophic situation.

(2) has a fuel capacity not exceeding 5 U.S. gallons

(3) is not capable of more than 55 knots calibrated airspeed at full power in level flight, and

(4) has a power-off stall speed which does not exceed 24 knots calibrated airspeed.

§ 103.3 **Inspection requirements.**

(a) Any person operating an ultralight vehicle under this part shall, upon request, allow the Administrator, or his designee, to inspect the vehicle to determine the applicability of this part.

(b) The pilot or operator of an ultralight vehicle must, upon request of the Administrator, furnish satisfactory evidence that the vehicle is subject only to the provisions of this part.

§ 103.5 **Waivers.**

No person may conduct operations that require a deviation from this part except under a written waiver issued by the Administrator.

§ 103.7 **Certification and registration.**

(a) Notwithstanding any other section pertaining to certification of aircraft or their parts or equipment, ultralight vehicles and their component parts and equipment are not required to meet the airworthiness certification standards specified for aircraft or to have certificates of airworthiness.

(b) Notwithstanding any other section pertaining to airman certification, operators of ultralight vehicles are not required to meet any aeronautical knowledge, age, or experience requirements to operate those vehicles or to have airman or medical certificates.

(c) Notwithstanding any other section pertaining to registration and marking of aircraft, ultralight vehicles are not required to be registered or to bear markings of any type.

Subpart B — Operating Rules

§ 103.9 **Hazardous operations.**

(a) No person may operate any ultralight vehicle in a manner that creates a hazard to other persons or property.

(b) No person may allow an object to be dropped from an ultralight vehicle if such action creates a hazard to other persons or property.

§ 103.11 **Daylight operations.**

(a) No person may operate an ultralight vehicle except between the hours of sunrise and sunset.

(b) Notwithstanding paragraph (a) of this section, ultralight vehicles may be operated during the twilight periods 30 minutes before official sunrise and 30 minutes after official sunset or, in Alaska, during the period of civil twilight as defined in the Air Almanac, if:

(1) the vehicle is equipped with an operating anticollision light visible for at least 3 statute miles, and

(2) all operations are conducted in uncontrolled airspace.

§ 103.13 Operation near aircraft; Right-of-way rules.

(a) Each person operating an ultralight vehicle shall maintain vigilance so as to see and avoid aircraft and shall yield the right-of-way to all aircraft.

(b) No person may operate an ultralight vehicle in a manner that creates a collision hazard with respect to any aircraft.

(c) Powered ultralights shall yield the right-of-way to unpowered ultralights.

§ 103.15 Operations over congested areas.

No person may operate an ultralight vehicle over any congested area of a city, town, or settlement, or over any open air assembly of persons.

§ 103.17 Operations in certain airspace.

No person may operate an ultralight vehicle within an airport traffic area, control zone, terminal control area, or positive control area unless that person has prior authorization from the air traffic control facility having jurisdiction over that airspace.

§ 103.19 Operations in prohibited or restricted areas.

No person may operate an ultralight vehicle in prohibited or restricted areas unless that person has permission from the using or controlling agency, as appropriate.

§ 103.21 Visual reference with the surface.

No person may operate an ultralight vehicle except by visual reference with the surface.

§ 103.23 Flight visibility and cloud clearance requirements.

No person may operate an ultralight vehicle when the flight visibility or distance from clouds is less than that in the accompanying table, as appropriate. (See Figure 1)

(Secs. 307, 313(a), 601(a), 602, and 603, Federal Aviation Act of 1958 (49 U.S.C. §§ 1348, 1354(a), 1421(a), 1422, and 1423; sec. 6(c), Department of Transportation Act (49 U.S.C. § 1655(c)).

Note: The FAA has determined that this regulation is not a major rule under Executive Order 12291. Because the rule will regulate a new user segment and because of substantial public interest, it has been determined that it is a significant rule pursuant to the Dept. of Transportation Regulatory Policies and Procedures (44 FR 11034; February 26, 1979).

The total projected costs of this rule may be found in a copy of the regulatory evaluation contained in the public docket. A copy of that evaluation may be obtained by contacting the person identified above under the caption "For Further Information Contact." It is certified under the criteria of the Regulatory Flexibility Act that this rule will not have a significant economic impact on a substantial number of small entities. These are very small entities involved in ultralight vehicle activities and the majority of those will be unaffected by the implementation of this rule.

Issued in Washington, DC, on July 30, 1982.

J. Lynn Helms
Administrator

COMMENTARY
THE RULE
Subpart A — General
§ 103.1 Applicability
(proposed § 101.1(a) (3)).

This section defines the term "ultralight vehicle." The proposed rule would have limited the term to single-occupant designs weighing less than 155 pounds, with a fuel capacity of 15 pounds or less, and which had no U.S. or foreign airworthiness certificate.

The final rule expands the definition to differentiate between powered and unpowered ultralight vehicles. The 155-pound weight limitation has been retained for unpowered designs and is the only criterion for those vehicles. Those ultralights equipped with powerplants must weigh less than 254 pounds empty weight.

In addition, powered ultralight vehicles must have a fuel capacity not exceeding five U.S. gallons and be incapable of more than 55 knots calibrated airspeed at full power in level flight. The power-off stall speed of a powered ultralight must not exceed 24 knots calibrated airspeed.

The rule restricts both powered and unpowered vehicles to single occupants and requires that the aircraft be used exclusively for sport or recreational purposes.

The FAA estimates that nearly all unpowered vehicles currently on the market will fall within the definition of ultralight vehicle. The new criteria will exclude approximately seven percent of the powered vehicle designs currently being marketed as ultralights, although many of those may be suitable for modifications to bring them within the scope of the definition.

Unpowered ultralight vehicles

A number of commenters, including the U.S. Hang Gliding Assn. (USHGA), object to the inclusion of "pure" hang gliders in the same definition as powered hang gliders. They raise the point that there are a number of distinctive operational differences between a pure hang glider and a powered vehicle which should be considered when assessing the necessity for regulations for these vehicles. The USHGA emphasizes its own self-regulation program and safety record.

The FAA recognizes that the measures taken by the USHGA to promote safety at USHGA launch sites have been effective, particularly those measures taken to protect the participants. However, the basic rationale for issuance of this rule is the safety of all users of the national airspace, not just the ultralight operators. The great majority of hang gliding operations will not be affected by these regulations because, as a number of commenters indicate, they are usually conducted in rural or remote areas, at low altitudes, away from areas where safety of other persons in the air or on the ground is compromised. It is only in congested areas, airport traffic areas and other areas frequented by aircraft involved in air commerce that these rules would restrict operations of unpowered ultralight vehicles.

The USHGA's self-regulation program lacks the legal authority to enforce requirements to ensure the safety of others. There is no requirement for any hang glider operator to be a member of the USHGA. Current hang glider publications have carried a number of articles describing hang glider operations which violate Part 91 regulations as well as the recommendations of Advisory Circular No. 60-10.

Those descriptions have included operations near and into clouds, low-altitude operations over open-air assemblies of persons and flights in close proximity to airports with large concentrations of airline and general aviation aircraft operations. Those potentially hazardous operations created the requirement for Federal regulatory limitations on hang gliders.

The proposed maximum weight restriction of less than 155 pounds was retained for unpowered ultralight vehicles to: (1) recognize the unpowered vehicles as a separate entity from those that are powered; and (2) ensure that the unpowered vehicles continue to meet essentially the same criteria that prevented their being classified as conventional gliders. Under this rule, those unpowered vehicles weighing 155 pounds or more must be certificated under the appropriate FARs. No specific comments were received which objected to the 155-pound limitation on unpowered vehicles.

Powered ultralight vehicles

A large number of commenters request that the proposed maximum empty weight of 155 pounds be raised for powered ultralight vehicles. The suggestions range from 180 to 350 pounds. The reasons offered include greater structural integrity, more opportunity for design innovations and the fact that many of the vehicles presently operated exhibit all of the other characteristics generally attributed to ultralights but weigh more than the proposed weight limit.

The FAA, by review of ultralight advertisements as of March, 1982, has concluded that the empty weights of most of those vehicles range from 150 to 250 pounds. It was further concluded that the higher weights resulted from improvements which provide greater structural integrity, better stability, more positive controllability and other safety-oriented additions which do not derogate the characteristics commonly associated with ultralight operations. Those characteristics are identified as low forward speeds, low wing loadings, low stall speeds, short takeoff and landing capability and no enclosures around the pilot.

Some commenters suggest that limitations of 220 pounds or 330 pounds be adopted because they are "international standards." This is not correct. Canada, England and Australia adopted 220 pounds as the maximum weight for a particular category of aircraft. In those countries, even if the weight limitation is met, the aircraft must be certificated and the pilots licensed. The 330-pound limit was established by the Federation Aeronautique Internationale for a category called "microlight aircraft." That category was established merely for the purpose of recording performance achievements of a particular group of aircraft.

The FAA agrees that the weight limitation for powered ultralight vehicles should be raised from the proposed 155 pounds. The 254-pound limitation was established because it closely corresponds to commenters' recommendations that the weight limitation be raised to at least 115 kilos, and because the vast majority of current vehicles on the market weigh less than 254 pounds. This weight does not include floats or safety devices intended for deployment in an emergency situation, e.g., parachutes and the harnesses and ballistic package necessary for deployment.

A large number of commenters recognize that, if the weight were raised, some restriction would have to be imposed to ensure that the characteristics associated with ultralights would be preserved. Those commenters include organizations such as the Experimental Aircraft Assn. (EAA), the Aircraft Owners and Pilots Assn. (AOPA) and the Powered Ultralight Manufacturers Assn. (PUMA).

The restrictions they propose range from simple wing loading values to complex aerodynamic formulas. They include maximum wing loading suggestions, minimum wing areas in relation to weight, maximum power capabilities in relation to weight and calculations of launch mass. Some commenters suggest, and the FAA considered, that the pilot be required to be exposed fully to the relative wind. This requirement was dropped to accommodate cold weather operations and to avoid stifling design and efficiency improvements within the parameters of an ultralight vehicle.

The maximum forward airspeed limitation was selected by the FAA because it is faster than almost all ultralight vehicles currently being sold but still places those vehicles in a significantly slower performance category than conventional aircraft. The determination and enforcement of this speed limitation is within the capability and resources of the FAA under the inspection requirements of the rule.

A number of commenters suggest maximum stall speed restrictions ranging from 18 to 25 mph, believing that this limitation would continue to ensure the safe nature of ultralight vehicles. The FAA believes that the ability of those vehicles to operate from surfaces other than those designed for aircraft is a factor which lessens the potential for collisions and reduces the interference with aircraft operations. A relatively slow stall speed is a major contributing factor in allowing ultralight pilots to operate in a safe manner.

A maximum power-off stall speed of 24 knots was chosen because it encompasses most of the vehicles currently on the market. The stall speed is easily determined through a simple calculation using information which is readily available to the FAA inspector when inspecting a specific vehicle.

The total allowable fuel capacity was raised from the proposed 15 pounds to 5 U.S. gallons. The decision to increase the volume of fuel is a direct result of the desire by the FAA, in response to public comments, to ensure that adequate fuel reserves are available for safe flight.

Single occupant

The rule limits both powered and unpowered ultralight vehicles to a single occupant. A few commenters suggest that two-seat versions be available for carrying passengers or for training purposes. The basis for allowing ultralight vehicles to operate under special rules which do not require pilot and aircraft certification is the "sport" aspect of the operation. For example, the assumption can be made that a person who elects, without pilot qualifications, to operate an uncertificated vehicle alone is fully aware of the risks involved.

This assumption does not hold true of a passenger selected randomly from the general public. Persons in the general public will likely assume that the operator has certificated pilot qualifications. Because pilot qualifications are not controlled or monitored, the single-occupant requirement is a necessary component in the continuation of the policies which allow the operation of ultralight vehicles free from many of the restrictions imposed on aircraft. Persons wishing to operate two-place vehicles have the availability of existing provisions of the FAR's for conducting such operations.

Recreation or sport purposes only

Recent activities and advertisements in ultralight-oriented publications (included in the docket) imply that commercial operations may be conducted by an uncertificated pilot in an ultralight which has not been certificated as an aircraft. Those types of operations are not allowed under the rule.

Several commenters suggest that ultralight vehicles be limited to sport or recreational purposes only. The position of the FAA has consistently been that these vehicles may be operated for sport and recreation purposes only. The justification for allowing the operation of these vehicles without requiring aircraft and pilot certification has been that this activity is a "sport" generally conducted away from concentrations of population and aircraft operations. Like any sport, the participants are viewed as taking personal risks which do not affect others not involved in the activity.

§ 103.3 Inspection requirements
(proposed § 101.55).

This section ensures the FAA's authority to inspect ultralight vehicles for compliance with the limits specified in § 103.1 and is retained in the final rule as proposed in Notice No. 81-6.

A large number of commenters object to the inspection requirements, believing that considerable FAA manpower and resources would be required in this effort. The USHGA and its membership contributed a majority of the objecting comments, citing the remoteness of hang gliding sites as impractical for the FAA to monitor.

Given the current level of ultralight activity, the FAA is confident that enforcement of the provisions of Part 103 can be accomplished with the existing resources. As is the case today, many investigations of suspected violations are prompted by reports received from pilots, air traffic controllers, citizens and other sources. The FAA foresees no appreciable increase in the number of these reports as a result of this rule.

§ 103.5 Waivers

In proposing to include ultralight operations under Part 101, ultralights would have been eligible for the waiver provisions applicable to all operations under that part. By removing the ultralight proposal from Part 101, the waiver eligibility for ultralights would have been lost. The FAA has concluded that the ultralight industry and the public would be best served by retention of waiver eligibility for these vehicles.

Thus, § 103.5 is added to the final rule, giving the ultralight operator the opportunity to apply for a certificate of waiver from any provisions of Part 103.

§ 103.7 Certification and registration

The intent of the FAA is to provide for safety in the national airspace with a minimum amount of regulation. Accordingly, those vehicles which meet the definition of "ultralight vehicle" will be exempt from FAA certification and registration requirements. Similarly, pilots of ultralight vehicles, as defined in this part, will not be required to possess FAA pilot certificates or airman medical certificates.

While this rule does not, at this time, require airman/aircraft certification of vehicle registration and is premised on the absolute minimum regulation necessary to ensure safety in the public interest, a continuation of burgeoning growth of the ultralight population could necessitate further regulation. The best practices and methods to preclude the need for further Federal regulation appear to at least include: self-regulation and self-policing, safety standards, membership in organizations and associations equipped to function and operate programs approved by the FAA, markings and identification of vehicles, programs including provisions similar to Federal Aviation Regulations relating to aircraft (both operation and airworthiness), etc.

FAA will continue to monitor performance of the ultralight community in terms of safety statistics, growth trends and maturity and, if indicated, will take additional regulatory actions to preclude degradation of safety to the general public while allowing maximum freedom for ultralight operation. In summary, it should be emphasized that the individual ultralight operator's support and compliance with national self-regulation programs is essential to the FAA's continued policy of allowing industry self regulation in these areas.

Pilot certification

A large number of commenters believe that there should be some requirement that pilots of ultralights be required to exhibit some knowledge and/or experience before being allowed to operate these vehicles. The suggestions range from no requirements to pilot certification under the requirements of Part 61.

The general groupings of the comments are: (1) no certification; (2) required ground training on regulations and conventional aircraft operations; (3) required ground training and instructor sign-off for unsupervised solo operations; (4) successful passage of a written test, such as the FAA glider pilot written examination; (5) issuance of an Ultralight Pilot Certificate by the FAA based on satisfactory completion of an examination and observed performance as the pilot of an ultralight; and (6) conforming to the certification requirements of Part 61 for student and private pilots.

The FAA endorses the ultralight community's efforts to develop and administer, under FAA guidelines, a national pilot certification program. At this time, however, pilots of ultralight vehicles are not required by Federal regulation to be certificated.

Aircraft registration

Some commenters, primarily state and local governments, recommend that these vehicles be registered and be required to display their registration number. The reasons center around identification of any offenders. The FAA's experience in identification of offenders and processing enforcement action validates their recommendations. The FAA endorses the ultralight community's efforts to develop and maintain, under FAA guidelines, a national registration system which would be immediately accessible to the FAA. However, registration of ultralight vehicles will not be required by Federal regulation at this time.

Aircraft certification

There are a small number of commenters who recommend additional Federal regulations requiring certification of ultralight vehicles to some design standards. The FAA has consistently refrained from the certification of these vehicles because they were operated by a single occupant for sport or recreational purposes. This policy is in accord with Federal regulatory policies regarding other sport activities. The pilots of these vehicles accept the responsibility for assuring their personal safety much as the drive of a moped street vehicle or a scuba diver does when engaged in his sport.

The FAA has noted and commends the efforts of the USHGA to establish design standards and flight testing of new hang glider designs. The FAA endorses the development of similar standards and testing of new powered designs by the ultralight community. However, the FAA presently has no intent to require certification of these vehicles by Federal regulation.

Subpart B — Operating Rules
§ 103.9 Hazardous operations
(proposed § 101.7).

This section prohibits any ultralight operator from engaging in activity which jeopardizes the safety of persons or property on the ground or in the air. The prohibition against hazardous flight or dropping of objects is common to the regulations pertaining to civil aircraft, and the FAA is addressing ultralight operations with equivalent stringency.

§ 103.11 Daylight operations
(proposed § 101.43).

The proposed rule would have limited the operation of ultralights to the hours between official sunrise and official sunset. The limitation on daytime operations was retained with an added provision for twilight operations under certain conditions. Other nighttime operations are not allowed.

A large number of commenters request that flight during the twilight periods of the day be allowed since those are prime times to conduct ultralight operations. They state that meteorological conditions are often best during those periods and are characterized by a lack of wind and turbulence.

The AOPA believes that calm air is particularly important for the novice flyer and provides an increased safety factor, especially during training when confidence building is essential. Many commenters believe that the available light is generally adequate to allow operations during those periods and that other craft could be safely avoided.

There are some commenters who believe that operations in Alaska should be excluded from the daylight operations section. They allude to the uniqueness of their "normal" day and how ultralight operations would be adversely affected.

Several comments support the original proposal and do not want operations during the nighttime hours. The primary concern centers around the difficulty in seeing these vehicles, especially at the higher altitudes, and the perceived inability of these operations to be conducted safely.

The FAA has observed ultralight operations during the twilight periods and has found the light available for such operations to be adequate in many instances. Operators were able to maneuver safely to avoid each other and also effect safe takeoffs and landings. Since most vehicles are operated at nearly the same altitude, they could be easily seen silhouetted against the lighted sky.

Operations were conducted in relatively close proximity to each other, and each operator was readily aware of the others' presence. The mild weather conditions which generally prevailed during the twilight periods combined with the controllability and maneuverability of these vehicles to enhance the safety factor for flight.

The FAA is concerned, however, that unlimited operations of this type could pose a threat to aircraft which operate at higher speeds and higher altitudes. The number of potential encounters between aircraft and ultralights increases significantly as ultralights operate into areas normally traversed by certificated aircraft. Also, the ability of aircraft pilots descending into the lower altitudes to see ultralights would be minimal due to the darkened backdrop of the ground. Pilots would often not be aware of such operations taking place and could easily overrun an ultralight without ever having visual contact.

The FAA has adopted an alternative which provides an acceptable level of safety to aircraft while still allowing ultralights to operate in uncontrolled airspace during this period of the day. The FAA's conclusion of this issue is to disallow ultralight operations in controlled airspace during the period from sunset to sunrise. This affords aircraft operators the margin of safety to which they are entitled and, at the same time, leaves adequate airspace to the ultralight operator during a 30-minute twilight period.

The FAA has determined that the occasional aircraft operation in uncontrolled airspace during the twilight period should not entirely preclude ultralight operations. The visibility from above of ultralights operating at very low levels can be significantly enhanced by the addition of an anticollision light on these vehicles. Such a light would provide the descending aircraft pilot with a distinct indication of the ultralight's presence. Additionally, it would enable ultralight operators to better see and avoid each other.

For the purposes of ultralight operation, an anticollision light is defined as any flashing or stroboscopic device that is of sufficient intensity so as to be visible for at least three statute miles. This regulatory approach does not impose on the ultralight owner the economic burden associated with a certificated lighting system. The ultralight must remain in uncontrolled airspace, and the anti-collision light must be operating during the twilight periods whenever the vehicle is in motion.

With respect to twilight operations in Alaska, the FAA recognizes that the periods of twilight are significantly different from those experienced in the lower latitudes. A review of the Air Almanac reveals that, in the upper latitudes, some days have no daylight periods but have over four hours of civil twilight. Civil twilight is defined as the period between official sunset and sunrise when the sun is less than six degrees below the horizon.

Regulations currently exist in Parts 91 and 101 which acknowledge the need to grant special allowances for operations in Alaska after sunset, and the FAA has determined that ultralights are entitled to the same consideration. Therefore, a provision to permit ultralight operations in Alaska during civil twilight has been added to § 103.11. The requirement to have an operating anticollision light during twilight operations is applicable to operations during this period in Alaska.

§ 103.13 Operations near aircraft and other ultralight vehicles; Right-of-way rules (proposed § 101.49).

The proposed regulations with respect to ultralight vehicle right-of-way are adopted. An additional provision is added to clarify the right-of-way

The Ultralight Rules - FAR Part 103

requirements in situations involving powered and unpowered ultralight vehicles.

The comments regarding right-of-way range from those who believe that unpowered ultralight vehicles should have the right-of-way over all other vehicles and aircraft to those who believe that the requirements of § 91.67 should be adopted, with unpowered ultralights being grouped with gliders and the powered ultralights grouped with airplanes. The most salient reasons cited include lack of maneuvering ability and inability to change location in the air quickly.

The suggestions and associated rationale do not reveal any areas which had not been considered during the formulation of the NPRM. The FAA has determined that uncertificated sport operations should not be given the right-of-way over all other aircraft. The small size and sport nature of the operations is a major factor in that determination.

It is unlikely that the pilot of aircraft will be able to see the ultralight vehicle as readily as the pilot of the ultralight vehicle will be able to or hear the larger aircraft. Due to the forward speeds of the majority of aircraft, it may be impossible for the aircraft to make sudden changes of direction required to avoid small objects sighted at close quarters. The FAA recommends that operators engaged in ultralight operations avoid, if possible, areas where significant operations of aircraft are occurring so as to minimize the risk of midair collisions.

Some ultralight operators express concern that if they are not given the right-of-way over aircraft, the pilots of those aircraft might deliberately fly in close proximity to the ultralights. In situations where this act can be substantiated, an investigation will be initiated to determine whether the pilot of the conventional aircraft operated in a careless or reckless manner in violation of § 91.9.

Some commenters recommend the establishment of areas where ultralight operations could be conducted and all aircraft operations would be prohibited. While the FAA has undertaken to identify locations on aeronautical charts where a specialized aeronautical activity, such as parachute jumping or gliding, is being conducted, no action is anticipated which would restrict other types of aeronautical activities in those areas and, similarly, no such action is contemplated for ultralights.

§ 103.15 Operations over congested areas (proposed § 101.47).

The proposed prohibition of ultralight vehicle operations over congested areas is retained in the final rule. The comments favoring an easing of the proposed rule focus on three main areas: (1) those who favor permitting operations with a minimum altitude ranging from 1000 to 3000 feet AGL; (2) those requesting that the minimum altitude requirements of § 91.79 be allowed; and (3) those who believe that no minimum altitude should be specified, especially for unpowered vehicles, due to the short field landing ability and small size of the vehicles.

The representatives of cities and towns who commented generally favor the prohibition, believing that uncertificated aviation activities have no place over congested areas.

The FAA's position is based on the fact that ultralight vehicles are not certificated as airworthy by any approved method and are flown by uncertificated pilots for sport or recreational purposes only. Similar limitations apply to the operations of experimental and restricted category aircraft based on catastrophic incidents which have occurred in the past.

The potential for such an incident makes the general issuance of the suggested authorization unacceptable. The FAA believes that concentrations of the general public must be protected from the possible dangers inherent in the operations of vehicles of uncertificated, possibly unproven designs. In specific limited instances, with appropriate operational limitations, ultralight operations may be approved over congested areas, through the waiver provisions of § 103.5.

§ 103.17 Operations in certain airspace (proposed § 101.45).

The NPRM proposed to require the ultralight operator to obtain authorization prior to operating within airport traffic areas, control zones, terminal control areas, and positive controlled airspace.

Operators of aircraft commented that the speed and visibility of ultralights are incompatible with other operations and that they should not be allowed at all in those areas. Some even suggest that a maximum operating altitude, such as 3000 feet AGL, be imposed on all ultralight operations.

The FAA shares the concern expressed by pilots who are wary of the ability to intermix faster aircraft safely with the relatively slow ultralights; but experience has shown that aircraft of significantly different performance characteristics can be accommodated when operations are conducted in accordance with specific authorizations.

There is considerable precedence in the form of glider operations, hot air ballooning and parachuting being conducted while aircraft safely transit the area. Historically, the greatest danger comes not from performance variables, but from operations unknown to the pilot or controller. The requirement to gain authorization before entering these airspace areas enhances the safety to all airspace users. The FAA has concluded that ultralight vehicles in compliance with the provisions of § 103.19 will be able to operate safely in those airspace areas.

Although the subject was not addressed in the NPRM, some commenters voice concern about ultralight operations conducted at or near uncontrolled airports, with many persons noting a need to develop standard operating procedures. The FAA agrees with the need to establish a compatible method of operation at uncontrolled airports but believes that the variables associated with each locality (terrain, runway configuration, and the physical properties of the airport) combine in such a manner to preclude a generalized nationwide regulatory approach.

The FAA has concluded that such operations could be handled much more efficiently by airport managers developing local procedures in concert with the ultralight community. In this way the available facilities can be used to the full extent while operational safety is maintained. Additionally, the interaction of the ultralight operators and the airport managers will serve as a basis for mutual understanding of the role this growing segment of aviation will play in the years ahead. The FAA encourages and supports efforts to reach such agreements and has been working with user groups in the development of guidelines for ultralight operations at uncontrolled airports.

§ 103.19 Operations in prohibited or restricted areas.

In this NPRM, requirements for operations of ultralights were included under the provisions of § 101.5.

In the final rule, the requirement for ultralight operators to obtain authorization prior to operating in prohibited or restricted areas is retained and restated under § 103.21.

Prohibited areas have been developed to provide for the safety and security of operations being conducted and to segregate activities considered to be hazardous to nonparticipating aircraft. Such operations in these areas include military and presidential security, flight training and testing, experimental weapons testing, and the launch and recovery of rocket-powered vehicles.

Many commenters recognize the need to limit access to these operating areas and accept the requirement to obtain permission prior to operating in these areas. A few commenters believe that this restriction should not apply to them and that ultralight vehicles should be allowed to operate at their own risk.

The FAA has determined that allowing any aeronautical activity to enter prohibited or restricted areas without prior authorization would derogate the purpose for which these areas were established. Avoidance of such areas by ultralight operators is not viewed as imposing a significant burden on ultralight operations.

§ 103.21 Visual reference to the surface
(proposed § 101.51).

NPRM No. 81-6 proposed that ultralight operators be required to maintain visual reference to the surface during all flight operations. This would ensure that the operator of an ultralight would have the opportunity to descend and land safely at any time without entering obscuring weather phenomena.

Many commenters support the proposal as reasonable and representative of normal ultralight operations. They recognize the possibility of being caught "on top" and the danger, both to themselves and to other airspace users, of trying to descend through a layer of clouds. A few commenters believe that visual reference to the surface is necessary only while climbing or descending and not while in level flight.

The FAA has determined that visual reference with the surface is necessary at all times. Experience with certificated aircraft has shown that many pilots, with fully instrumented aircraft, have been caught "on top" and required assistance from Air Traffic Control to descend safely. Flying "on top" or between cloud layers often presents visual illusions which cannot be verified without instrumentation.

The effect of these illusions is to disorient the airman spatially, with a resulting loss of control of the craft. It takes a well-trained and disciplined pilot to ignore what information the human senses are providing and rely on the instrumentation aboard the aircraft.

In the case of ultralights, there is relatively little, if any, instrumentation with which to confirm the flight attitude of the vehicle. Further, if the ultralight operator should get caught "on top," there is no alternative available but to descend unannounced through the clouds. The ultralight operator would be risking not only his own life, but the lives of persons who rely on the safeguards inherent in certificated aviation.

The FAA has determined that inclusion in the final rule of the requirement to maintain visual reference with the surface is necessary to reduce the potential for collisions and ensure the safe operation of ultralight vehicles.

§ 103.23 Flight visibility and cloud clearance requirements
(proposed § 101.53).

The flight visibility and cloud clearance requirements proposed in the NPRM are the same as those under § 91.105, the basic minima for VFR flight operations by fixed-wing aircraft. Since ultralight vehicles will be sharing the same airspace, the FAA has determined it is practical to apply the same operating minima.

Many commenters to this proposal are receptive to the similarity in visibility requirements for all airspace users. Many ultralight operators indicate an appreciation for the inherent safety in being able to see and avoid obstructions and other aeronautical activities. Establishment of specific visibility standards is viewed as enhancing the legitimacy and the utility of ultralight operations.

Some commenters believed that the distance from clouds should be reduced to "clear of clouds." Their basis for such a change centers around the difficulty in determining actual distances from clouds.

Other commenters suggest that hang gliders be allowed to continue their practice of operating near and in the base of clouds. Their rationale is based on the added lift available from being in close proximity to cumulus clouds. Some hang glider operators fear that the restriction on in-cloud operations would eliminate their ability to vie for long-distance and high-altitude records.

The FAA cannot support the operation of ultralights in or near clouds. A specific distance from clouds is required when operating in controlled airspace, primarily due to the presence of aircraft conducting instrument flight operations through the clouds. The cloud clearance requirements serve as a practical buffer to reduce the possibility of having an aircraft exit the clouds on an unalterable collision course.

Operating too close to clouds does, in effect, cause a blind side in the aviator's vision. Operation in and near clouds severely restricts the ultralight operator's ability to see and avoid, an ability that is paramount in allowing ultralight operations to take place.

In maintaining a safe distance from clouds, the FAA has concluded that ultralight operators can reasonably approximate, when operations are being conducted, the required distance from clouds. Experience with other segments of aviation has shown that it is readily apparent that, when operations approach an unsafe distance from clouds and adherence to the prescribed minimum distance determination becomes relatively easy. Therefore, retention of the flight visibility and clouds clearance requirements, as proposed, is essential for maintaining airspace safety.

"VFR conditions must be maintained at all times by ultralight vehicles in flight. This requires minimum distances from clouds in different kinds of airspace, as well as minimum visibility and constant reference to the surface."

A-3
Clarification of FAR 103

OPERATIONS OF
ULTRALIGHT VEHICLES
Q & A

The Federal Aviation Administration recently sent to all Flight Standards Division managers a series of questions and answers it designed to clarify the meaning and intent of Federal Aviation Regulation Part 103 (those rules applying to ultralights), as well as to help flight standards inspectors respond to queries from ultralight operators. The Air Safety Foundation and the FAA culled through the list and compiled and edited the following to provide operators with insight into the agency's interpretation of the Part 103 rules.

S 103.1(a) - Single Occupant Only

1. Q. Can a two-place aircraft that also meets the weight and speed definition of an ultralight be operated by a single occupant who is not a certified pilot?

A. No. In order for a vehicle to be operated under FAR 103 it must strictly meet the definition in 103.1 that is to be used and *intended to be used* for manned operations in the air *by a single* occupant and the vehicle shall not have a U.S. or foreign airworthiness certificate issued to that serial numbered vehicle.

2. Q. If a person is receiving instruction in a two-place aircraft for the purpose of qualifying to operate an ultralight vehicle, must that instruction be given by a certified flight instructor?

   A. No. The check-out instruction that one pilot gives another for the purpose of qualifying to fly an ultralight may be given by any pilot qualified to fly the two-place aircraft with passengers. This would not be instruction creditable towards any FAA certificate unless the instruction were given by an authorized CFI.

3. Q. Can a waiver be issued to allow flight instruction for compensation in a two-place aircraft that has an experimental certificate and operating limitations that do not include the purpose of flight instruction?

   A. No. Applicants for such relief should be advised that no waiver will be issued. Applicants may however seek an exemption if they can meet the requirements of Section 11.25 of the FAR.

S 103.1(b) - Sport or Recreation Only

1. Q. Can a waiver be issued to allow for the commercial operation of an ultralight?

   A. No. FAR 103.1(b) specifically states that the operation must be for sport or recreational purposes *only*. The basis for the relaxation of the operating rules allowing the operation of the ultralight vehicle without airworthiness certification, registration, or pilot certification is the fact that these vehicles would only be used for sport or recreation. No commercial operations will be authorized.

2. Q. Can ultralights be used for aerial application?

   A. No. An ultralight vehicle may only be operated for the purpose of sport or recreation. FAR Part 137, Agricultural Aircraft Operations, requires that any person who conducts agricultural operations must hold an agricultural operator's certificate issued under FAR 137. Only certificated aircraft including certain experimentally certificated aircraft may be used under Part 137.

3. Q. Can an ultralight be used to tow advertising banners like conventional airplanes?

   A. No. See answer to question #1. An experimentally certificated aircraft may have banner towing included in its operation limitations; however, experimental aircraft may not be used for commercial purposes or to earn compensation.

4. Q. Do small hang balloons (a one-man, basketless balloon) fall under the applicability of Part 103?

   A. Yes. If the balloon, harness, and tank weight is less than 155 pounds,

Clarification of FAR 103   193

they are considered unpowered ultralights and can be operated under the applicability of S 103.1.

5. Q. Can an ultralight be operated as a public aircraft?

A. Yes. If a government declares its intention to operate a particular craft as a public aircraft then it must be registered under Part 45 and must comply with the operating rules of 91 rather than 103. There is no such thing as a "public vehicle" recognized in the FAA ACT.

S 103.1(e) - Powered

1. Q. What are the "devices which are intended for deployment in catastrophic situations" that are excluded from the empty weight of a powered ultralight?

A. Parachutes and their deployment devices or other yet to be invented lifting devices which would allow a gentle landing after a catastrophic failure, are considered to be such devices. Safety devices such as seat belts, roll cage and wheel brakes, are considered part of the airframe and should be included in the empty weight.

2. Q. Stall speed varies according to gross weight of the aircraft. What weight payload should be used to determine the stall speed for qualification for the power off 24 knot stall speed required to operate under FAR 103.

A. A standard occupant weight of 170 pounds would be acceptable in making the determination for a particular craft.

S 103.3 - Inspection Requirements

1. Q. What documentation must an ultralight operator present to show he/she is in compliance with FAR 103?

A. The rule is operational and does not require any specific documentation. Each vehicle must meet the applicability requirement in order to operate under the rule. Operators must find it helpful to carry a letter or other sort of documentation certifying to the fact that his particular serial numbered machine meets the applicability of FAR 103.1 as various local officials may ask for documentation. A *registration including specification* issued by a recognized ultralight organization such as the ultralight division of AOPA or EAA will also be useful.

S 103.7 - Pilot Certification

1. Q. If an airman possesses a pilot certificate and is operating under FAR 103 but violates a section of that rule, will action be taken against that certificate?

A. Qualified yes. An airman is responsible to conduct all operations in accordance with all applicable FARs. Each case will, of course, be judged on the merits, but the fact that the airman is operating in noncompliance does make possible a sanction against his airman certificate. Operating an ultralight does not exempt an airman from possible certificate action for noncompliance.

2. **Q.** When a certificated airman is operating an ultralight vehicle under FAR 103, must he/she carry their pilot certificate in accordance with FAR 61.3?

   A. No. It is not required for operations under FAR 103. (It will be of benefit to the airman to carry his/her certificate as evidence of their pilot competency and knowledge of the rules for any local law enforcement officials or others who may require identification and evidence of competence. Many airport officials are requiring some documentary evidence of operator competence before operations may be conducted on their airport.)

S 103.15 - Congested Areas/Minimum Altitudes

1. **Q.** Will a waiver be issued to operate over congested areas of cities and towns?

   A. In order to issue a waiver the FAA must be assured that an equivalent level of safety exists and that the operation is in the public interest. At the present time we cannot conceive of a situation that would warrant a waiver of this section.

2. **Q.** At what altitudes may an ultralight vehicle operate over people and structures on the surface?

   A. Section 103.15 prohibits the operation of an ultralight over any congested area which is similar to the prohibition applied to experimental aircraft. Thought should be given to the ability of the craft to make an emergency landing without injury to persons and property on the ground. Operations over groups of persons or structures would be contrary to Section 103.15. Operations so low or close as to cause a hazard to persons and property are contrary to Section 103.9. No other specific minimum safe altitudes are defined in the rule. Due to the unreliable nature of the uncertificated power plants, all operation of ultralights should be conducted at such an altitude that would allow a safe emergency landing. Operations contrary to this would be in violation of Section 103.9.

Enforcement/Accident

1. **Q.** What action should be taken if an ultralight is found not to be in compliance with the applicability section, Part 103?

   A. The ultralight must meet the applicability requirements of 103.1 in order to be privileged to operate under the relaxed rules of FAR 103. If the craft does not meet the applicability, then the operator must comply with all rules relative to an aircraft. The operator should be informed of this in writing and a record kept of this notification complete with an identifying description of the vehicle; i.e., color, model, serial number, engine model and any other helpful identifiers.

Clarification of FAR 103 195

If the craft is observed in flight, treat the case as you would any other operation of an uncertificated aircraft. Pursue enforcement action as appropriate.

2. **Q.** Must accidents in ultralight aircraft be reported to the NTSB and investigated in accordance with Part 830 of the NTSB rules?

   A. No accident report will be requested nor will an accident investigation be accomplished. If an operator wishes to report an accident, refer him to the NTSB where he may forward his report. Do investigate acts of noncompliance resulting in accidents and, as always, take appropriate action.

QUESTIONS CONCERNING THE OPERATION AND HANDLING OF ULTRALIGHT VEHICLES

In the airport environment:

Q. May ultralights use both controlled and uncontrolled airports?

A. Yes, provided the airport manager does not expressly prohibit ultralight operations. Airports which have Federal funding should accommodate ultralight activities, but only to the extent that they do not derrogate safety.

Q. Can an ultralight operate on a portion of the airport other than the established runways?

A. Yes, with the airport manager's consent. We encourage the segregation of ultralights and aircraft due to the different operating characteristics of each.

Q. What about noise restrictions on such vehicles?

A. The FAA has not set noise standards for ultralight vehicles under FAR 36. In the interest of being good neighbors, ultralights should be operated so as to have a minimum noise impact on the community.

In the airspace:

Q. Do ultralight operators have to get ATC authorization to operate anywhere in an airport traffic area, even though they will not be using the primary airport?

A. Yes. FAR Part 103 requires ATC authorization within an airport traffic area, control zone, terminal control area, or positive control area. Authorization may be via telephone or radio on a case by case basis, or may be contained in a letter of agreement with an ultralight flight park. In such cases, the airport/flight park manager should be a party to such agreements.

Q. Can ultralights use normal VHF aviation radio frequencies to contact ATC?

A. Yes, provided they have the proper license from FCC for an aircraft radio. Additionally, the FCC license directs the use of an FCC call sign by the ultralight operator. Each license has the following statement printed on it: "Identify transmissions by FCC control numbers." For instance, if the FCC control number 12834U, then the ultralight operator should use that identification in the transmission. Example: "Ultralight 12834U."

Q. Can ultralights operate in a MOA?

A. Yes. MOAs are not restricted for the purpose of VFR flight operations. Ultralight operators should be encouraged to avoid such areas when they are in use due to the nature of military training taking place within these areas.

Q. Can ultralights legally operate in a VFR flight corridor through a TCA?

A. Yes. There are no restrictions, nor minimum certificate requirements to operate in a VFR environment. Authorization is required only in the areas specified in the rule, ATAs, CAs, TCAs, PCAs. However, ultralight operators should be operating through VFR corridors. This may lead to an aircraft overtaking the ultralight with relatively little room to maneuver to avoid a collision. Also, ultralights should be discouraged from coming in close proximity to any major airport area, especially those with TCAs.

Q. Will ultralight sites be depicted on sectional charts?

A. No decision has been made to include ultralight operating areas on the sectional charts. Areas of significant ultralight activity will be considered as candidates for charting.

Air Safety Foundation Guidelines for the Operation of Ultralight Aircraft at Existing Airports with Diagrams

GUIDELINES FOR ASSESSING THE IMPACT OF JOINT CONVENTIONAL/ULTRALIGHT AIRCRAFT OPERATIONS OF AN AIRPORT

The prospect of ultralight operations from an established airport should be viewed by the airport operator as an opportunity rather than a problem. The operator of a publicly-owned airport must consider that ultralight owners are also local taxpayers and a broader base of airport-using taxpayers is an asset. Operators of private airports must also weigh the benefits of a broader base of users should the airport be threatened by non-users.

But not every established airport can accommodate a joint-use arrangement safely while at the same time maintaining a good relationship with airport neighbors. Therefore, the following should be considered.

- **Liability** - Normal insurance coverage may exclude ultralight operations since ultralight aircraft and pilots are not required to possess FAA certificates. Special provisions for these operations are available at little or no additional cost, however.
- **Traffic Density** - Ultralight operations conducted from a dedicated portion of the airport will not impact on airport capacity or safety. Ultralight operations from the normal runway can be safety conducted jointly with local conventionally certificated aircraft provided that all users adhere to guidelines and procedures established by the airport operator. Joint operations from a runway by either transient conventional or transient ultralight pilots should, if possible, be accommodated by creation of a prominently marked, dedicated ultralight operating area. In all circumstances, ultralight pilots will prefer a dedicated takeoff/landing area for convenience and safety.
- **Local Winds** - Safe ultralight operations require relatively calm, steady wind conditions. Terrain or buildings on or near the airport which tend to produce eddies must be considered carefully.
- **Population Areas** - Current production ultralights are quieter than many conventional aircraft. But ultralight operations tend to be conducted in early morning and late evening calm. Also, ultralights are flown at lower altitudes and lower speeds. The impact on nearby population areas must be assessed with those characteristics of the sport in mind.

The Air Safety Foundation staff is available to provide guidance to airport operators on questions of operations and safety. We have also developed posters to assist in making both ultralight and conventional aircraft pilots aware of the operating practices followed by the other.

AOPA Air Safety Foundation
7315 Wisconsin Avenue
Bethesda, Maryland 20814

GUIDELINES FOR THE OPERATION OF ULTRALIGHT AIRCRAFT AT EXISTING AIRPORTS

The guidelines that follow should be used to form specific operating rules and traffic patterns at individual airports. The real key to implementing these procedures is wide dissemination to all interested parties. Every effort should be made to ensure that both ultralight and conventional aircraft pilots are aware of these special procedures.

ASF Guidelines for the Operation of Ultralights at Existing Airports 199

Ultralight Operations where a Separate Operating Area is Available

Figure 1

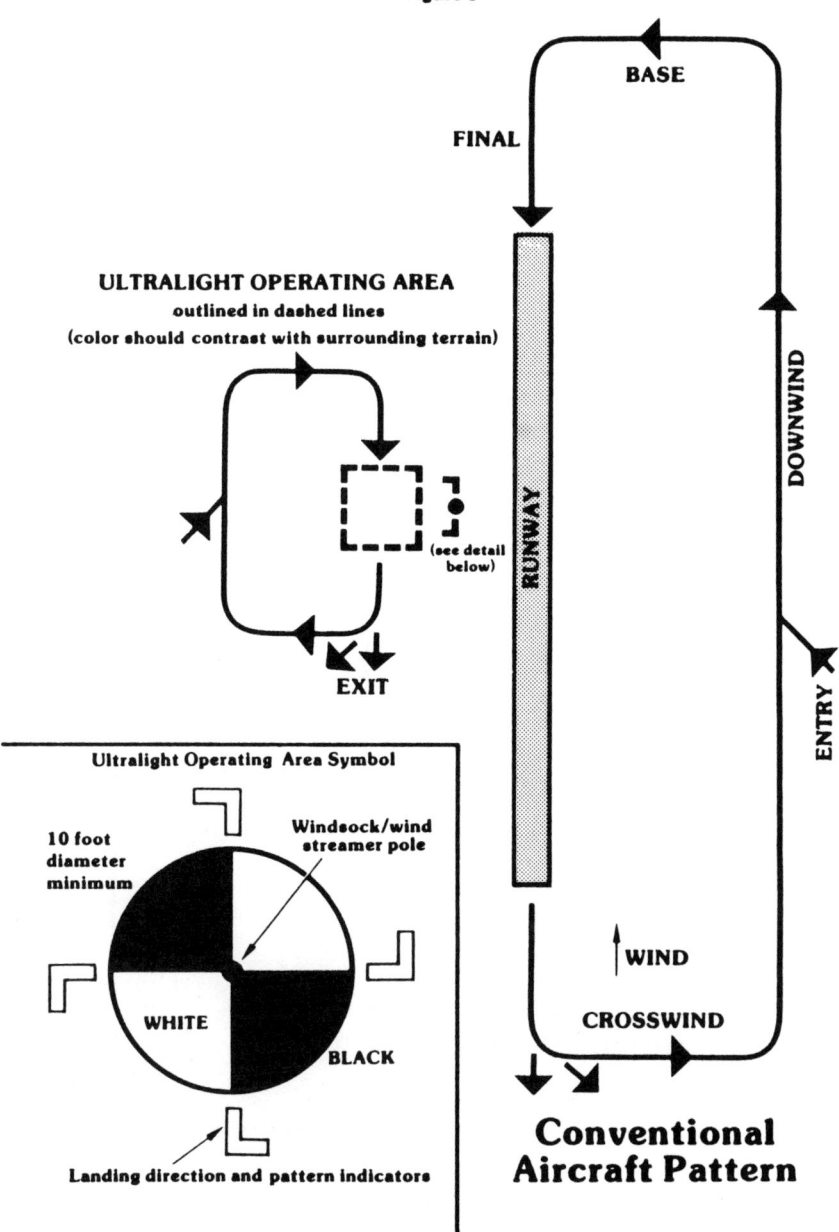

Fig. A-3. Diagram illustrating an ultralight operating area located next to a conventional aircraft pattern.

Simultaneous Ultralight and Conventional Aircraft Operations from a Single Runway

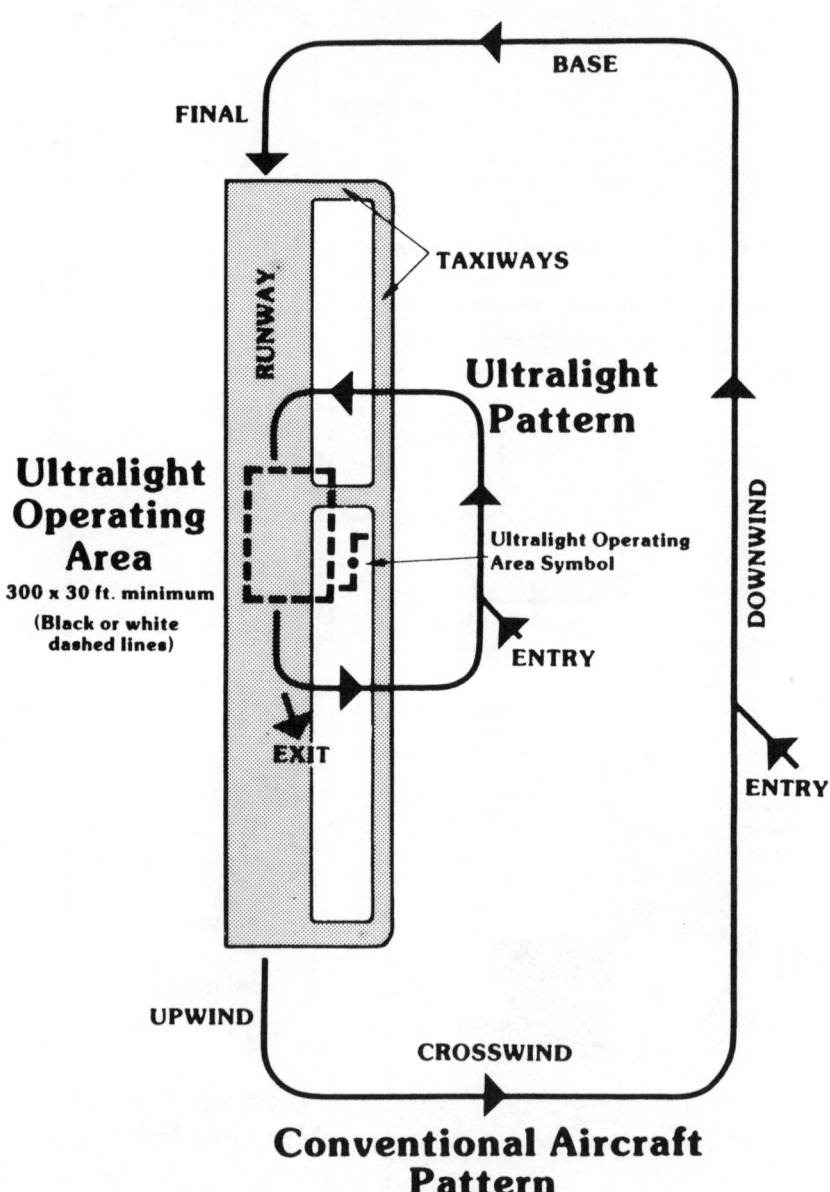

Fig. A-4. Diagram illustrating how ultralight and conventional aircraft patterns can operate from a single runway.

ASF Guidelines for the Operation of Ultralights at Existing Airports

I. AIRPORT SURFACE OPERATIONS
  A. Dedicated Takeoff and Landing Area
  1. If at all feasible, powered ultralights should operate from a dedicated area of the airport.
  2. That area should be clearly marked for both ultralight and conventionally certificated pilots with the distinct symbol shown in figure 1. Note that the segments indicate the direction of the ultralight pattern. This Ultralight Operating Area symbol is an ideal place to put a windsock or windstreamer for the benefit of the ultralight pilot. Wind direction and velocity are of great safety importance. (This marker should be permanent for airports with a great amount of ultralight activity, but may also be made of movable panels.)
  3. The ultralight takeoff and landing area should be a square at least 300 feet on a side, or a circle with a radius of 150 feet, plus adequate clearways for initial climb and approaches.
  4. The dedicated area should not encroach on an active runway closer than 300 feet from the centerline of that runway.

  B. Simultaneous Operations From a Single Runway
  1. When it is necessary for powered ultralights to operate from the runway in use by conventionally certificated aircraft, a segment of that runway should be designated for that purpose. The designated segment should result in the shortest possible runway occupancy by ultralight aircraft, consistent with their normal and safe taxi speeds (figure 2).
  2. When ultralight and certificated aircraft are jointly using a runway, a knowledgeable observer should be located at the ultralight entry point to that runway to assist ultralight pilots in timing operations safely with respect to highspeed traffic. If frequent operations by high performance aircraft (approach speeds of 80 knots or more) may be expected on a runway in joint use, and if the airport has Unicom, it is recommended that the observer be equipped with a receiver tuned to the Unicom frequency. This procedure may be waived for single ultralight operations, particularly for runways on which the traffic volume is very low.

  C. Ground Support Activities
  1. The airport operator should specify which areas of the airport may be used for the movement and parking of automobiles and other surface vehicles used in support of ultralight movements.
  2. Airport management should make policy on the admission of children, pets and non-flying observers to the ultralight area and insure that the policy is known and understood by every person admitted.
  3. Hours of activity should be specified by airport management consistent with the airport's good neighbor policy.

II. FLIGHT OPERATIONS
A. Traffic Patterns
1. In joint operations, the powered ultralight traffic pattern should have the same general rectangular configuration described in the Airman's Information Manual.
2. The ultralight pattern should be smaller than that of conventional aircraft.
3. The ultralight pattern should be 500 feet below the standard pattern altitude for the airport unless otherwise dictated for safety or noise abatement purposes.
4. When jointly using a runway, both conventional aircraft and ultralight patterns should be on the same side of the runway with the ultralight pattern inside and below that of conventional aircraft.
5. When ultralights are operated from a discrete area of the airport, the ultralight pattern should be adjusted to avoid crossovers of the runway in use by conventional aircraft.
6. Care should be taken to ensure that patterns are oriented so that the ultralight aircraft will not cross an active ramp area or taxiway at low altitude.
7. The ultralight pattern and recommended routes to and from the airport should be designed so that in the event of a loss of power, the aircraft will be able to make a safe power-off landing without undue hazard to either the ultralight or public property. For example, high density housing areas, schools and large bodies of water should be avoided.

B. Pattern Exit and Entry
1. In general, ultralight aircraft should exit and enter the ultralight pattern in conformity to the guidelines in the Airman's Information Manual.
2. If terrain or populated areas dictate non-standard traffic pattern exit-entry procedures, the airport management should specify what flight paths will be followed and ensure that ultralight pilots understand them.
3. Ultralight pilots desiring to enter or depart the pattern across a runway in use by conventional aircraft should cross the runway at ultralight pattern altitude after insuring there will be no conflict with conventional aircraft. A specific crossing point may be designated if desired.

C. Airport management should specify all ultralight flight paths and altitudes in the vicinity of the airport to insure minimum noise impact on airport neighbors.

III. OPERATING ANNOUNCEMENTS
A. An Ultralight Operating Area symbol should be displayed permanently at each airport where ultralights operate on a regular basis.

ASF Guidelines for the Operation of Ultralights at Existing Airports

Ultralight operations should be included in Unicom airport information.
C. Each unusual concentration of ultralight activity — such as a competitive event — should be included in NOTAMS.

IV. SAFETY

A. Ultralight pilots should demonstrate to airport management a knowledge of appropriate airspace regulations and the airport operating guidelines in the Airman's Information Manual. (All persons, whether they are FAA certificated airmen or not, are responsible for compliance with Federal Air Regulations.)
B. Before operating from an airport, each ultralight pilot should be briefed on airport policy, traffic pattern procedures in various wind conditions, population areas to be avoided, local weather phenomena and area terrain features significant to ultralight safety.
C. Ultralight pilots should be familiar with local IFR procedures and the non-standard patterns flown by aircraft operating IFR.
D. Ultralight and conventional aircraft pilots alike should be aware of the effect of wake and helicopter rotor turbulence on ultralight aircraft and the separation criteria to be observed.
E. The first solo flight of an ultralight pilot should be conducted only under the supervision of an experienced ultralight instructor who has taken appropriate precautions to insure there will be no conflict with other traffic.

V. REFERENCES

Federal Air Regulations
   Part 91 - General operating and Flight Rules
   Part 101 - Moored balloons, Kites, Ultralight Vehicles, Unmanned Rockets and Unmanned Free Balloons

FAA Airman's Information Manual

FAA Advisory Circular 90-66 - Recommended Standard Traffic Patterns for Airplane Operations at Uncontrolled Airports

"Before starting any engine, be sure the wheels are chocked, or the aircraft is otherwise restrained from movement. The propeller area is to be clear of spectators, as well."

Air Safety Foundation Ultralight Safety Tips

Flying ultralight aircraft can be one of the most enjoyable ways of pursuing sport aviation. As with any sport, there are certain practices that make it easier and safer. The tips that follow are based on the experience of Air Safety Foundation staff and inputs from the training departments of ultralight manufacturers. They will make flying ultralights more fun for everyone.

PERSONAL
- Obtain flight training from a qualified instructor who is experienced in your type aircraft.
- Ensure that you are mentally and physically fit to fly.
- Always wear a helmet.
- Always wear a parachute when flying at an altitude that will permit its deployment.
- Be familiar with the basics of aerodynamics, rules of the air (Federal Air Regulations), meteorology, engine operation and maintenance procedures.

AIRCRAFT
- Each aircraft should be maintained in accordance with the manufacturer's maintenance schedule.
- A thorough preflight check should be conducted prior to **each** flight.
- Stay in contact with the manufacturer's customer service department to ensure you know **all** of the changes that affect the safety of your aircraft.
- Never make a change or modification that could reduce the structural strength or alter the flying qualities of any aircraft without first checking with the manufacturer or other qualified person.
- All repairs should be of high quality, using accepted aircraft repair practices. (FAA Advisory Circular 43.13-1A-Acceptable Methods, Techniques, and Practices: Aircraft Inspection and Repair contains many such practices.)
- Have a dealer or other experienced person take a close look at your aircraft once a year (more frequently if you fly often) to see if there are any defects you may have overlooked.
- Thoroughly inspect all parts of the aircraft any time you have a hard landing, drag a wingtip or any other similar incident.
- Treat your aircraft as if your life depended on it — it does!

FLYING SITE
- Should have clear approaches, free from obstructions.
- Buildings, lines of trees and other obstacles should not be positioned so as to cause mechanical turbulence in the landing/takeoff area.
- Definite operating procedures should be developed and made available to all pilots at the site.
- If the site is an airport:
  - Find out whether or not ultralights are welcome at the field.
  - Become familiar with all local procedures and adhere to them.
  - Be particularly alert for conventional aircraft patterns and conventional aircraft pilots who may not be familiar with local ultralight procedures.
  - Be very cautious concerning wake turbulence effects from conventional aircraft.

OPERATIONS
- Never let an inexperienced person fly your aircraft, whether they are a pilot or not.
- Never start your engine when there is a crowd around the aircraft.
- Be aware of the wind direction and velocity at all times.
- Don't make steep turns, abrupt pitch changes or any radical maneuver below 300 feet above the ground.
- Always have a place to land in case of engine failure.
- Maintain sufficient clearance from mountains and hills to avoid the effects of turbulence often associated with these areas.

- Be aware of the noise your aircraft produces — prolonged exposure to it can be quite annoying to people on the ground.
- Don't fly low over persons, vehicles or buildings — you can impress people on the ground better than you can in the air.
- Practice engine-out landings regularly — it helps keep surprises to a minimum.
- Report all ultralight incidents to the Air Safety Foundation — turn a mishap into a learning experience.
- Fly as if your life depended on it!

WEATHER

- Be familiar with local weather conditions and trends.
- Always obtain a current weather forecast prior to flying.
- Learn as much as you can about winds, turbulence, shears and gradients both for the atmosphere in general and the local area in particular. These are the most troublesome aspects of ultralight flying.
- Unless you are a very experienced ultralight pilot, never fly in winds above 10-12 knots, or when the winds are gusty.
- Be alert for sudden changes in the weather, both on the ground and inflight.

The Air Safety Foundation has been in existence for over thirty years as a non-profit, tax exempt foundation dedicated to aviation safety. Each year the ASF conducts training courses for more than 20,000 pilots of conventional aircraft and reaches an additional 15,000 pilots through their free Operational Seminars. The ASF staff also conducts on-going research into general aviation safety problems at a Flight and Technology Laboratory located in Frederick, Maryland.

With this broad aviation safety background, the Air Safety Foundation has become deeply involved in the ultralight movement, providing the same services for it as it has for all other aspects of general aviation. The ASF has been instrumental in bringing together manufacturers, governmental agencies and individuals interested in making ultralights more enjoyable and safe. A major product of these meetings has been national guidelines for ultralight pilots, aircraft and flight parks.

These guidelines are being expanded to include airman certification, aircraft registration, accident reporting and airworthiness standards, which will be administered by the Foundation in the interest of promoting the sport and fostering safe practices.

For more information concerning these programs and others for ultralights, write to:

AOPA Air Safety Foundation
Ultralight Programs
7315 Wisconsin Avenue
Bethesda, Maryland 20814

"... It is recommended you take dual flight instruction in a two-place ultralight. It's the best way to learn how to fly."

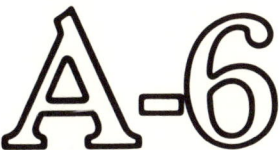

Air Safety Foundation Ultralight Incident/Accident Reporting Form

AOPA AIR SAFETY FOUNDATION
Ultralight Incident/Accident Reporting Form

The information you provide in this form will be used by the Air Safety Foundation to analyze structural, training and technique problems within the sport. The results will be compiled and published periodically in safety bulletins and in *Ultralight Pilot;* however, your name and the names of the people involved in the incident/accident will not be made public. Please remember that your attention to detail may help others avoid problems. After you complete the form (please type or print legibly), fold it and mail it; no postage is necessary.

_____ Incident—minor structural damage or near miss

_____ Accident—injuries or structural failure

Information on Person Piloting the Aircraft

Name _____ Sex Male ( )   Female ( )

Address _____

_____ Age _____

Phone_____Weight _____

Had pilot received any instruction in an ultralight?   Yes ( )   No ( )

   If yes, number of hours _____

Was aircraft operator registered with ASF?   Yes ( )   No ( )

Total ultralight hours of pilot _____

Total years of ultralight experience_____

Length of time since previous flight _____

Physical and emotional condition of pilot at time of occurrence _____

_____

Other flying history (include FAA certificates, if any)_____

_____

Ultralight Information

Type of ultralight: Manufacturer _____

Model _____ Serial No._____

Homebuilt? Yes ( )   No ( )

Type of engine _____

Modifications to ultralight_____

_____

Did pilot have: Airspeed indicator?   Yes ( )   No ( )
                      Parachute?   Yes ( )   No ( )
                      Helmet?   Yes ( )   No ( )

Number of hours logged on ultralight by above pilot _____

Total number of hours logged on above ultralight_____

Did pilot receive an owner's manual?   Yes ( )   No ( )

Site Information

Location (be exact) _____

_____

_____

Date of occurrence_____

Time of day_____

Describe lighting: Daytime ( )   Sunset/Sunrise ( )   Dark ( )

Wind conditions: Velocity_____ Direction_____

Turbulence: None ( )   Mild ( )   Severe ( )

Purpose of flight: Cross-country ( )   Local ( )
                   Instructional ( )   Demonstration ( )

ASF Ultralight Incident/Accident Reporting Form

Occurrence

Describe the incident/accident, including diagrams if needed. Please include any information you feel is important to the understanding of this occurrence. Use additional sheets if necessary.

Are photographs available?   Yes ( )   No ( )

Do you feel this occurrence could have been avoided (if so, how)? _____
_____
_____

Injuries to pilot: Fatal ( )   Major ( )   Minor ( )
Hospitalization required?   Yes ( )   No ( )
Damage to ultralight: Destroyed ( )   Substantial ( )   Minor ( )   None ( )

List specific parts damaged _____
_____
_____

Please provide the following information in case we need to contact you for clarification of points contained in this report.

Name _____

Address _____
_____

Phone _____ Date _____

AOPA Air Safety Foundation
Box 865
Frederick, Maryland 21701

"Fly only in calm or light wind conditions. Avoid gusty conditions like the plague."

Ultralight Vehicle Airman Certification Program

CONCEPT

The AOPA Air Safety Foundation Ultralight Airman Certification Program is centered around a written, oral and flight proficiency test that is based on a set of minimum standards developed by a group concerned with the safe and responsible growth of the sport. Ultralight Pilot and Flight Instructor certificates will be issued to those who meet certain experience requirements and pass all tests successfully. The Ultralight Flight Instructor will act as the Air Safety Foundation's designated examiner in the field to test and issue temporary certificates. A central computer data base of all airman transactions will be maintained at Foundation headquarters and will be available for information requests from government agencies and other legitimate organizations on need to know, non-commercial basis.

This data base will be used to notify ultralight airmen of free operational seminars given by the Foundation. Additionally, airman exposure data gained during the certification and renewal process will serve as a statistical base from which accident and growth rate data can be compiled.

Each registered ultralight airman will receive a quarterly safety newsletter at no charge.

CERTIFICATION PLAN

A. This program will be conducted by the AOPA Air Safety Foundation, 7315 Wisconsin Avenue, Bethesda, Maryland, 20814.
B. The person responsible for this program will be Mr. John J. Sheehan, Director of Ultralight Programs, Telephone: 301-951-3973.
C. Administration and conduct of the program:
   1. The Air Safety Foundation (ASF) is a non-profit, tax exempt corporation that has no members. The Foundation currently conducts aviation training on a national basis which will provide a structure for this program. This program will be made available to all persons without regard to organizational affiliation.
   2. A network of ASF designated Ultralight Flight Instructors (UFI) will be authorized to issue ultralight airman certificates. These instructors will meet the requirements for UFI as listed in subsequent paragraphs. A UFI who has held his certificate for less than one year will be unable to issue a certificate to a UFI applicant.
   3. Prerequisites necessary to take competency tests:
      a. Ultralight Pilot - minimum age - 16 years
      b. Ultralight Flight Instructor
         1. Minimum age - 18 years
         2. At least 20 hours of flight time in ultralights over a minimum period of six months. The six month requirement is unnecessary if the applicant holds at least an FAA private pilot certificate.
         3. At least five hours of flight time in each type of aircraft in which he instructs.
         4. Flights from at least three different ultralight airports or sites.
         5. Five landings within the 30 days prior to the proficiency demonstration.
         6. Completion of an approved instructor training course within one year prior to application for certification.

Grandfathering - For each applicant for UFI certification who applies to July 1, 1983, the experience, proficiency, and approved course requirements will be waived upon endorsement from a recognized manufacturer, dealer or other responsible organization which certifies that the requirements have been met.

   4. Test Content. The written test must be successfully completed prior to taking the oral or flight portions.
      a. Written - all tests will be multiple choice or true/false type.

1. Pilot - 50 questions covering the following subjects:
   FAR Part 103
   Applicable portions of the Basic Airman's Information Manual
   Airspace
   Meteorology
   Aeronautical chart interpretation
   Basic aerodynamics
   Aircraft performance
   Maintenance and repair
   Engine operation
   Accident reporting
2. Flight Instructor - 75 questions covering the following subjects:
   All knowledge areas required of the ultralight pilot (See the following section)
   The learning process
   Elements of effective teaching
   Student evaluation and testing
   Flight instruction techniques

   b. Flight competency/practical demonstration
      1. Pilot
         Preflight
         Engine start
         Taxi
         Takeoff
         Rectangular pattern
         Slow flight
         Landings
         Emergency procedures
      Grandfathering - Prior to July 1, 1983, the flight proficiency requirement may be waived at the discretion of the UFI.
      2. Flight Instructor
         Conduct a flight lesson
         Analysis and correction of student errors
         Performance of standard flight training procedures and maneuvers
      Grandfathering - The provisions listed at the end of item 3 apply.
   c. All applicants may be orally quizzed on any written or flight test item. Instructors will be instructed to conduct at least a thirty minute oral quiz with each applicant.
5. Standards
   a. Written test. Minimum passing grade for each written test will be 70%.
   b. Flight test/practical demonstration. Each operation called for will be individually judged by the examining UFI with regard to demonstrated good judgement and safety. Any individual opera-

tion executed in an unsafe manner or series of acts that demonstrate poor judgement shall be disqualifying. UFI applicants will be required to demonstrate a high level of competence and skill in the practical and knowledge aspects of their tests.
  c. Oral. Applicants should answer at least 70% of the questions correctly. Failure to answer any questions considered to be critical by the UFI shall be disqualifying.
  d. Retesting after failure. An applicant will be required to wait three days prior to retesting if any portion of the test process is not successfully completed.
6. Training. Ground training will be offered on a nationwide basis for initial pilot and instructor training and for recurrent instructor training. Since none of these courses are fully prepared at this time nor required prerequisites until July 1, 1983, they will not be addressed at this time.
7. System controls. Each UFI will be provided with a detailed flight test guide and with standardized written exams, both of which will be required in the conduct of competency tests. Peer pressure, public opinion and economic considerations will be the principal constraints put upon individual UFI's. Any report of malfeasance by any instructor will be fully investigated by the Foundation. Spot checks of instructors will be conducted on a regular basis.
8. Certificate contents. (Actual certificate pending legal approval) Pilot certificate (Front)

AOPA AIR SAFETY FOUNDATION

This certifies that

(Name)

(Address)

has demonstrated a minimum level of

understanding or skill in the elements

listed on the reverse side.

Expires (date)

Signature of holder   (/S/) Number

Reverse (Pilot)
Ultralight Aircraft Operations

Knowledge	**Skill**
FAR Part 103	Preflight
AIM (Applicable portions)	Start
Airspace	Taxi
Meteorology	Takeoff

Ultralight Vehicle Airman Certification Program

Aeronautical Charts
Aerodynamics
Aircraft Performance
Maintenance/Repair
Engine Operation
Accident Reporting

Rectangular Pattern
Slow Flight
Landing
Emergency Procedures

Reverse (UFI)
Ultralight Aircraft Operations

Knowledge

FAR Part 103
AIM (Applicable portions)
Airspace
Meteorology
Aeronautical Charts
Aerodynamics
Aircraft Performance
Maintenance/Repair
Engine Operation
Accident Reporting
Learning Process
Effective Teaching
Evaluation and Testing
Instruction Techniques

Skill

Preflight
Start
Taxi
Takeoff
Rectangular Pattern
Slow Flight
Landing
Emergency Procedures
Analysis and Correction of Errors
Training Procedures and Maneuvers

9. Termination and renewal. Each applicant will acknowledge a statement when they sign the application for a certificate that states, "The AOPA Air Safety Foundation has the right to revoke and recall this certificate for cause at any time". This will give a good measure of quality control even if the certificate is not returned upon request, because insurance companies, states, flight park operators and the FAA will be provided with lists of revoked certificates.

   Certificates will be valid for two years and will be renewable. Pilots may renew by contacting a UFI for a short refresher briefing which is a review of critical information and an update of new information. The UFI may only renew by attending an approved UFI refresher course.

10. Administrative processing. Examining UFI's will issue temporary certificates to applicants who successfully pass the applicable tests. The completed airman application form will be forwarded to ASF headquarters for verification and data base entry. The output from the data base update will be a computer produced certificate which will be mailed to the new airman.

    A fee (initially five dollars) will be charged to cover processing and safety newsletter mailings.

11. Record keeping. Current airman records will be maintained in an in-house computer for a minimum of 27 months. Expired and renewed records will be maintained on a microfiche for a minimum of five years. Records will be released to the FAA as needed; to individual state governments, but only records of residents of that state; and to law enforcement agencies on a case basis. Lists and/or mailing labels of airman records will not be sold.

    Any other FAA approved organizations engaging in ultralight airman certification must not sell or indiscriminately release information received from ASF for the purpose of administrative continuity.
12. Additional features. Each registered ultralight airman will be sent a free quarterly safety newsletter which will include an analysis of accident reports and accident prevention articles. The data base will be used to notify ultralight airmen of free operational seminars to be given by the Foundation on a nationwide basis. Airman exposure data gained during the certification and renewal process will serve as a statistical base from which accident and growth rate data can be compiled.

Ultralight Vehicle Registration Program

CONCEPT

The AOPA Air Safety Foundation Ultralight Vehicle Registration Program will provide ultralight aircraft owners with a method of theft prevention, proof of ownership, and of receiving safety and airworthiness information concerning their specific aircraft. Registration markings will be assigned by the Foundation to applicants who can show ownership. The registrant will be issued a set of tamper-resistant labels containing the registration number to mark critical parts of the aircraft, a wallet-size registration certificate and instructions on how the visible wing markings are to be positioned. The registration will be valid as long as the aircraft is not sold or destroyed, the owner dies or two years elapse, whichever comes first. Provisions are made for renewal and transfer.

A central computer data base of all registration transactions will be maintained at Foundation headquarters and will be available for information requests from government agencies and other legitimate organizations on a need to know, non-commercial basis. All transactions will be accomplished by mail.

This data base will be used to notify ultralight aircraft registrants of free operational seminars given by the Foundation. Additionally, aircraft data gained during the registration and renewal process will serve as a statistical base from which accident and growth rate data can be compiled.

Each registered ultralight owner will receive a quarterly safety newsletter at no charge.

REGISTRATION PLAN
A. This program will be conducted by the AOPA Air Safety Foundation, 7315 Wisconsin Avenue, Bethesda, Maryland, 20814.
B. The person responsible for this program will be Mr. John J. Sheehan, Director of Ultralight Programs, Telephone: 301-951-3973.
C. Administration and conduct of the program:
   1. This program will be made available on a national basis via advertising in a variety of trade publications and direct mail solicitation. This registration system will be made available to all persons without regard to organizational affiliation.
   2. All registration transactions will be accomplished via mail, with processing being done at Air Safety Foundation headquarters.
   3. Prerequisites. All applicants must possess an ultralight vehicle that meets the criteria stated in FAR Part 103 and be able to show proof of ownership.
      a. Proof of ownership. The applicant for registration of an aircraft must submit proof of ownership with this application. A copy of sale of aircraft or kit is usually sufficient proof. If the applicant did not purchase the aircraft from the last registered owner, the applicant must submit conveyances completing the chain of ownership from the last registered owner to the applicant. If the aircraft was built from parts (amateur-built), a notarized affidavit describing the aircraft and stating that it was built from parts should be submitted as proof of ownership instead of a bill of sale.
   4. Registration Numbers. Numbers will be randomly assigned unless a special number is requested. If a special number is desired, this number may not exceed five characters and may be one to five numbers (77777), one to four numbers and one suffix letter (7000A) or one to three numbers and two suffix letters (700AA). The letters "I" and "O" cannot be used due to possibility of confusing them with numbers. An additional fee is charged for requested numbers.
   5. Duration of registration. The aircraft registration is valid for two years or until one of the following occurs:
      1. Sale of the aircraft
      2. Destruction of the aircraft
      3. Death of registered owner

Ultralight Vehicle Registration Program 221

If any of these events occur, the registration is no longer valid and must be returned to the Air Safety Foundation and all registration markings removed from the aircraft.

Renewal is accomplished by returning the renewal form and aircraft activity report to the Foundation.

6. Registration Certificate content. (Final wording is awaiting legal approval).

AOPA AIR SAFETY FOUNDATION

Ultralight Aircraft Registration

Registration Mark (Marking)

Aircraft Type (Designation)	Engine Type (Designation)
Aircraft Serial No. (Number)	Engine Serial No. (Number)

Issued to: (Name)

           (Address)

Date of Issue: (Date)	Expiration Date: (Date)

This certificate is issued for registration purposes only and is not a certificate of title. The AOPA Air Safety Foundation does not determine rights of ownership as between private parties.

(LOGO)

ASF R-1 (8/82)

---

7. Visible marking requirements. Each registered aircraft will be required to display the registration marking as follows:
    a. Characters shall be of roman capital style and contrast in color with the background to which they are affixed.
    b. Character must be six inches high and four inches wide, except the number "1" which must be one inch wide and the letter "M" and "W" which may be six inches wide. Characters must be formed by solid lines one inch thick.
    c. The registration marking shall be permanently affixed to the underside of the lowest left wing. The marking shall be oriented spanwise with the top of the characters pointing forward. The center of the marking shall be at least three feet but not more than five feet from the left wing tip.
    d. Letters shall be spaced a minimum of 1½ inch apart.
    e. Additional registration markings may be applied to the aircraft as desired.
8. Administrative processing. Registration applications and associated proof of ownership will be verified and entered into a central computer data base at Foundation headquarters. The registration card will then be computer produced and mailed to the registrant.

9. Record keeping. Current registration records will be maintained in an in-house computer for a minimum of 27 months. Expired and renewed records will be maintained on microfiche for a minimum of five years. Records will be released to the FAA as needed; to individual state governments, but only records of residents of that state; and to law enforcement agencies on a case basis. Lists and/or mailing labels of registration records will not be sold.
    Any other FAA approved organizations engaging in ultralight registration must not sell or indiscriminately release information received from ASF for the purpose of administrative continuity.
10. Additional features. Each registered ultralight owner will be sent a free quarterly safety newsletter which will include an analysis of accident reports, airworthiness notices and accident prevention articles. The data base will be used to notify ultralight owners of free operational seminars to be given by the Foundation on a nationwide basis. Aircraft exposure data gained during the certification and renewal process will serve as a statistical base from which accident and growth rate data can be compiled.

Suggested Standards for Ultralight Flight Parks

BACKGROUND

The powered ultralight vehicle (ultralight) movement has grown rapidly since its inception to the point where it has become a significant factor in general aviation. While many ultralight operators have been successfully integrated into conventional airport operations, it is generally agreed that the ideal situation would be for them to have a dedicated flying site of their own. This will allow the ultralight movement to have the maximum opportunity for unrestricted operations which will enhance enjoyment of the sport and accommodate their unique flight training requirements. The term "flightpark" is being used to distinguish these areas from conventional airports.

Persons desiring to build flightparks and state aeronautical agencies have expressed an interest in establishing guidelines for these sites. In response to this need, the AOPA Air Safety Foundation, in conjunction with concerned state agencies, and ultralight manufacturers and operators, have formulated the guidelines listed below. These guidelines are the first look at the subject and may not address all of the applicable issues. Therefore, all interested parties are encouraged to communicate their ideas on the subject to the Air Safety Foundation.

The FAA has not yet issued an official policy on the subject. The Air Safety Foundation is maintaining liaison with them on the subject, however.

GUIDELINES

Definition: A flightpark is an area of land that is used or intended to be used for the landing and takeoff of powered ultralight vehicles.

Notification: Each person desiring to establish or deactivate a flightpark shall notify the appropriate state agency and the Federal Aviation Administration. It is recommended that FAA Form 7480-1 be used to notify both agencies.

Flightpark Classifications: The following classifications are not meant to be mandatory, rather they are intended to assist both operators and controlling agencies in site selection and usage.

A. Class I - Unlimited class. Suitable for all training operations and competition.
   1. A minimum level surface of 2,500 x 500 feet aligned within 40 degrees of the prevailing wind.
   2. Clear approaches to both ends of the primary runway. An obstruction clearance plane of 15:1 extending from either end of a minimum included landing surface of 1,000 feet long outward for a minimum of 2,000 feet and 150 feet wide is recommended for this application.
   3. Nearest controlled airport a minimum of five miles away and the nearest uncontrolled airport a minimum of two miles away.
   4. Significant residential areas at least two miles away.
   5. Favorable zoning rules.
B. Class II - Suitable for most training activities.
   1. A minimum level surface of 1,000 x 300 feet aligned within 40 degrees of the prevailing wind.
   2. Approach/departure paths should be free of tall obstructions.
   3. A minimum of one mile from residential areas.
C. Class III - Limited use. Not suitable for training. Intended for private activity only.
   1. A minimum surface of 300 x 100 feet aligned within 40 degrees of the prevailing wind.

General Design Considerations

A. Careful consideration must be given to obstacles of any nature that might cause significant wind-generated mechanical turbulence or wind shadow in the landing and takeoff area.

B. A fence or other barrier must be used to segregate the aircraft operating area from other areas used for the general public, and aircraft maintenance and administrative purposes.
C. A minimum of one wind direction and velocity indicator capable of indicating a wind velocity of a minimum of 5 mph should be installed adjacent to the runway.
D. An ultralight operating area symbol should be installed at the base of the wind indicator or other appropriate location.
E. That portion of the aircraft operating area protected by the obstruction clearance plane or at least a 200 x 100 foot rectangle designating the center of operating area should be clearly outlined by paint, powdered lime or markers.
F. Hazardous areas on the landing surface and significant obstructions on or adjoining the flightpark should be clearly marked.

Operations

Each flightpark operator should specify traffic patterns for both ground and flight operations. Noise sensitive areas should be detailed and noise abatement procedures specified, if necessary.

Accident Reporting

Each flightpark operator shall be required to report to the AOPA Air Safety Foundation any reportable aircraft accident which occurs on the flightpark or for an aircraft which originates at the flightpark. (A reportable accident is one in which a death occurs or an individual must be treated at a hospital or other medical facility, or where aircraft receives substantial damage.) This report must be made within seven days of the occurrence.

"Remember. Angle of attack is controlled by the elevator, which governs airspeed via the stick."

A-10
AOPA Ultralight Division Competition Rules

Copyright © 1983 Aircraft Owners and Pilots Association
(Reprinted by Permission of AOPA)

FOREWORD

Aircraft Owners and Pilots Association, the world's largest civil aviation organization, and its Ultralight Division have two goals in sanctioning and promoting powered ultralight competition. The first is to establish a nationwide system for choosing pilots who outperform their peers in a predetermined series of tasks by being best able to combine flying skills, knowledge of the aircraft, and understanding of weather conditions. The purpose of such a goal is to reward pilots who have worked to achieve excellence through precise and careful flying. Our second goal is to present powered ultralight flying in a visible, enjoyable, and educational manner for pilots and the public.

Through such competition manufacturers can better realize the needs of pilots and improve ultralight aircraft, pilots can compare their skills with those of others, and spectators can learn about the sport in a controlled environment. These rules are intended to guide both meet personnel and pilots in achieving the above stated goals.

We believe that **skillful flying** is synonymous with **safe flying.** Therefore, all competition should consist of events carefully planned to maximize safety and enjoyment.

GENERAL RULES

1. All persons participating in, or in any way connected with, an AOPA/UD sanctioned meet shall be bound by the rules of the competition.
2. Interpretation of the rules shall be made by the Meet Director in consultation with the primary officials (Chief Judge, Safety Director, Chief Scorer, etc.) when necessary. All judging calls are final.
3. Special rules may apply on a day-to-day basis at the discretion of the Meet Director for the purpose of accommodating certain site requirements and/or weather restrictions.
4. In the event of inclement weather, the Meet Director shall determine an alternate schedule and shall post it prior to the beginning of the competition.
5. The flight tasks shall be chosen by the Meet Director according to weather conditions and time factors.
6. At least two (2) of the sanctioned tasks must be offered to competitors.
7. At least five (5) rounds must be completed to qualify for competition points. The Meet Director may authorize additional rounds.
8. Unofficial flying and practice flying may be scheduled and controlled by the Meet Director.
9. All aircraft must comply with the FAA definition of powered ultralights. Contestants may be asked to demonstrate aircraft compliance with FAR Part 103.
10. An ultralight regarded by the Meet Director as not airworthy may be disqualified from competition.
11. Substitution of aircraft is not allowed. Sharing of aircraft among contestants is not allowed.
12. Pilots must be members of the AOPA Ultralight Division.
13. Pilots must comply with FAR Part 103 rules.
14. Pilot minimum requirements may be established by the Meet Director.
15. Prior to the opening of the contest, all pilots must demonstrate pilot competency and aircraft airworthiness by completing a qualifying circuit around the field pattern in the ultralight to be used in the meet.
16. A pilot regarded by the Meet Director as not competent may be disqualified from competition.
17. Only pilots who have attended the preflight briefing may participate in the selected task.
18. Departure procedures will be described at the preflight briefing. Each competitor must be in position at appointed time. Failure to appear results in loss of the heat.

19. Pilots must remain clear of designated runways until their appointed flying time.
20. Pilots must walk ultralights from one runway to the other or between competition areas unless officials declare otherwise.
21. Pilots must wear protective headgear while flying. Parachutes are recommended.
22. Pilot intentionally causing an inflight conflict with an opponent will receive a loss for the heat.
23. It is the pilot's responsibility to fly safely, to operate in accordance with all rules and regulations, and to fly in an unambiguous manner across course gates and around pylons.
24. Pilots who fly over spectator area or buildings are subject to disqualification from the meet.
25. Damage to the aircraft sustained while competing disqualifies the pilot for that heat. The meet will **not** be delayed to allow competitors to repair their craft. A pilot will **not** be allowed to fly a damaged ultralight except upon discretion of the Meet Director who may authorize an ultralight to be flown that possesses a bent landing gear. Pilots are responsible for pointing out landing gear damage to officials prior to flight to avoid disqualification.
26. In the event that a disabled aircraft occupies the normal landing area, use another landing area.
27. No profanity or harassment of officials is allowed. Penalty is removal from meet for unsportsmanlike behavior.
28. Pilots must not approach scoring table unless accompanied by Ombudsman.
29. If a competitor's engine fails to start, the two-man heat is put on hold for no more than ten (10) minutes. The other pilot may cut engine until the competitor's engine starts. Other heats may be started during this period. If engine is not started by the end of the ten (10) minutes, the pilot receives a loss for the heat.
30. Engine must not be started in spectator areas.
31. No illegal drugs or alcohol may be consumed eight (8) hours prior to flying.
32. Subject to the Meet Director's approval, non-U.S. citizens may be registered if they are members of the AOPA Ultralight Division, and if they possess an ultralight pilot rating or a pilot certificate.
33. Meet will end at designated time or earlier, at the discretion of the Meet Director.

OPERATIONAL STANDARDS
Administration

Matters relating to planning the competition, competitor and aircraft entry requirements, administration and conduct of the competition, and certification of results may exceed AOPA/UD sanction requirements.

Aircraft Inspection

All aircraft must be inspected by the pilot after assembly and before it is flown in competition. A log manual certifying an amount of minimum operation time may be required for each engine/aircraft combination.

Competition Point System

The purpose of the Competition Point System (CPS) is to provide a nationwide ranking for competition pilots. It also is a method by which at least twenty-one (21) pilots who have proven their ability to win in competition can be chosen to compete in the AOPA/UD National Convention Championship. To receive CPS points the pilot must compete in AOPA/UD sanctioned meets. Competition results are reported to AOPA headquarters by the Meet Director on the Competition Final Standings Report Form.

AOPA headquarters establishes a national ranking for competition pilots, by selecting each pilot's three (3) best meet scores of the flying year. The flying year is defined as the close of Labor Day holiday weekend to the close of Labor Day holiday weekend the following year.

Points

— One (1) point is awarded to each contestant who completes the contest, i.e. competes through to the end of the contest.
— Additional points are added to pilots completing the contest based on the number of pilots completing the contest divided by ten (10). A maximum of five (5) points per pilot is awarded under this provision.
— Additional points are awarded to the top ten (10) winners of the competition according to their standing as follows:

Additional Points

Place	Points
1	15
2	13
3	11
4	9
5	7
6	5
7	4
8	3
9	2
10	1

LANDING

A landing is defined as the point both main wheels of the aircraft come in contact with the ground. All landings must be within a designated area to constitute a scored flight except where otherwise authorized. Pilots must remove their aircraft from the runway immediately after landing. Pilots must land immediately upon receiving a designated signal due to emergency or production reasons.

LAUNCHING

A launch is defined as the point in time that an ultralight begins ground roll to compete in the required task. When ready for takeoff, pilots will signal to the Flight Line Director by giving thumbs up sign. Launches will begin with a takeoff immediately after the Flight Line Director gives the start signal. Pilots failing to apply throttle and effecting a takeoff when signaled by the Flight Line Director will receive a loss for the heat.

LAUNCH REFUSAL

A pilot may refuse to launch if the pilot believes conditions are unsafe. The refusing pilot and those in that heat move back two heats. If a pilot in both the following heats refuses launch, then the launch is closed until conditions change according to the judgement of the Flight Line or Safety Director. Three consecutive launch refusals closes the launch. If three refusals do not occur, the competitor who refused originally must launch when his/her turn comes again or receive a loss for the heat. Misuse of this option for other than unsafe flying conditions may result in disqualification from the task and a loss of the heat.

OFFICIALS

1. **Organizers** have responsibility for producing and financing the meet but do not supersede the Meet Director in questions of competition operation. The organizers appoint the Meet Director.
2. **Meet Director** has operational responsibility for the competition, its tasks, format, judging, scoring and logistics. The Meet Director appoints the Safety Director, Chief Judge, Chief Scorer and other operational officials.
3. **Safety Director** has final say on the suitability of conditions and equipment and arranges for medical aid in conjunction with the meet organizers. The Safety Director's authority should supersede and be independent of the Meet Director.
4. **Chief Judge** is responsible for appointing judging personnel, assigning operational equipment as required by the tasks, and scoring. The Meet Director should consult the Chief Judge on matters of pilot performance. The Meet Director and Chief Judge may be one and the same in a small meet.
5. **Chief Scorer** is responsible for all data and information relating to pilot scores and reports to the Chief Judge. The Chief Scorer computes and posts all standings and round positions.
6. **Other Officials**
    a. Pylon Judges — perform service as directed by the Chief Judge.
    b. Assistant Scorers — perform service as directed by the Chief Scorer.
    c. Flight Line Directors — perform service as directed by the Meet Director.

d. Ombudsman — serves as a liaison between competitors and officials. The position may be held by the Meet Director or other official.

PILOT BRIEFING
All pilots must attend the scheduled pilot briefings in order to compete in the day's tasks. Generally, weather and task requirements will be given at these briefings.

PILOT EQUIPMENT
Helmet must be worn while flying. Parachutes are recommended. Radios are not allowed unless specific permission is stated in the individual meet rules.

PROTESTS
All complaints should be registered with the Chief Judge for investigation. If the pilot is unsatisfied with the result, a written protest must be submitted (with an appropriate fee, if any) not later than one hour after said complaint. A written protest will be ruled on by the Protest Committee (normally Meet Director, Chief Judge and Chief Scorer). Judging calls may not be protested. (See General Rules 2.)

Scoring errors should be reported to the Chief Judge. The Chief Judge will direct the Chief Scorer to make any changes.

If the Meet Director determines that a reflight is required, it will take operational precedence over all flights from a later round. Reflights may be awarded due to unfair conditions, failure or judging equipment or irreconcilable judging errors. The protest fee is returned only if the Protest Committee rules in favor of the protesting pilot.

REGISTRATION
Membership
Pilots must be members of the AOPA Ultralight Division to qualify for registration in the competition. Pilots who are not members may qualify for filling out and signing the AOPA/UD membership application form and paying dues at registration. AOPA Ultralight Division dues are $29 per year. If the individual is a current member of AOPA, they may join the Ultralight Division for $15. This is stated on the membership application form. Pilots paying $15 membership dues must list their AOPA member number. All AOPA members are issued a member card that contains the AOPA member number and renewal date.

Pilots must be members of the AOPA **Ultralight Division** to compete in the contest. A nonmember pilot competing in the AOPA/UD sanctioned contest is a violation of the sanction agreement and can void the sanction, thereby voiding the results of the meet and points awarded in the national Competition Point System.

Contestant Master Roster

Each pilot must be listed on the contestant Master Roster. If the individual joins AOPA/UD at registration, the amount of dues paid by the pilot is to be listed on the right hand column of the roster reserved for this purpose. The column will be of help in totaling the amount of dues to be forwarded to AOPA headquarters at the close of registration. The roster may be used as the list of entrants distributed to pilots at the seeding meeting discussed in Section II, Seeding Competitors. It is helpful, therefore, if the roster is compiled legibly.

SCORING

1. Pilots will receive either a win or a loss for each heat.
2. First round will be matched by seeding.
3. In each successive round pilots will fly within their win/loss group unless an uneven number exists in that group.
4. Lower win/loss groups may be cut at the discretion of Meet Director for efficiency purposes.
5. Ties within a win/loss group at the scoring of the task will be resolved by using sanctioned tie breaking tasks.
6. Cumulative scoring of all round results is used to determine final placement of pilots in the meet.

SIGNALS

Flags or markers waved by the Flight Line Officials shall be the primary means of communicating during task operation. Launch, landing and emergency signals will be designated at the pilot's briefing.

Launch signal consists of waving a green flag vertically.

Do-not-land signal consists of waving a red flag horizontally.

Emergency signal consists of waving a yellow flag in a figure 8 pattern.

TASKS

All tasks will be conducted one-and-one except Climb and Duration which will be operated one-on-one, i.e. both contestants launching at the same time.

SANCTIONED TASKS
Accuracy Landing
Climb and Duration
Figure Eight Maneuver

AOPA SANCTIONED TASK
ACCURACY LANDING

ACCURACY LANDING
This event tests the pilot's skill in performing an accurate and a safe power-off landing.

General Description
A. Each flight consists of a takeoff, flight around the field pattern, and a landing. Meet Director will declare left or right pattern before the round.
B. The object of the event is to safely land as closely as possible to the landing line.

Procedure
A. Pilots position their aircraft for takeoff as directed. Second pilot should launch when first pilot is on final to reduce the advantage of knowing the first pilot's results.
B. Pilot commences takeoff run when signaled by the Flight Line Director and flies the pattern as briefed at the pre-flight meeting.
C. Downwind leg should be flown parallel to and at an appropriate distance from the runway.
D. Engine should be killed prior to the landing line on the downwind leg.
E. Rollout from base leg to final should be at no more than medium bank (25-35 degs.).
F. Rollout to final approach should be completed at no less than 50 feet above the ground.

Scoring
A. The pilot measured as landing closest to the line wins the heat. In this event a landing is defined as the point where both main wheels of the aircraft come in contact with the ground **with no further departure.**
B. 360 degree turns earn the pilot an automatic loss for the heat.

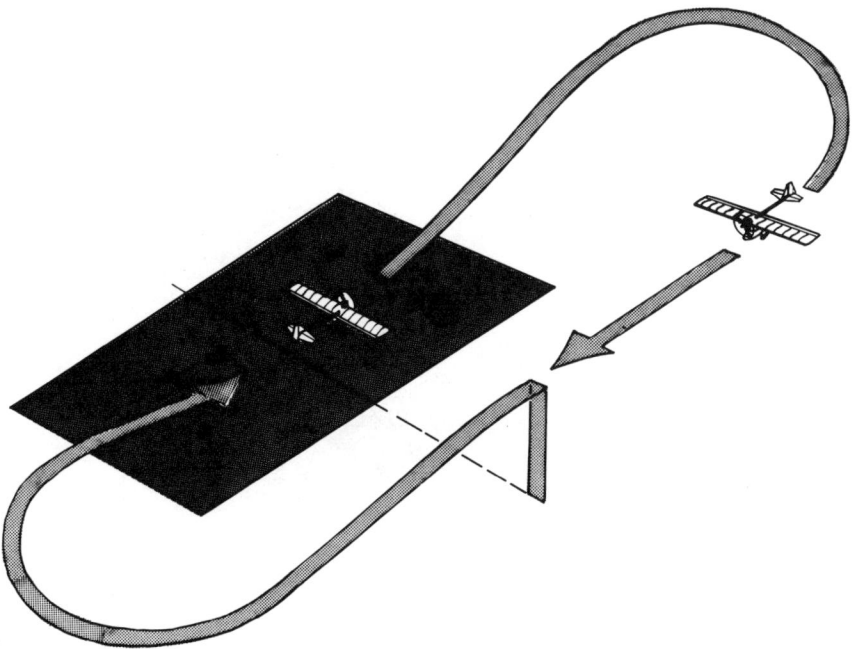

Fig. A-5. An illustration of the "Climb and Duration" event, with each ultralight allocated one quart of fuel. Longest aloft wins.

AOPA SANCTIONED TASK
CLIMB AND DURATION

CLIMB AND DURATION

This event tests the pilot's skill at soaring, and it provides an opportunity to demonstrate operational efficiencies of the aircraft.

General Description
A. The objective of the event is to land within the designated landing area later than one's opponent.
B. Each flight consists of a takeoff, duration maneuvering, and a landing.
C. The event will be flown one-on-one.

Procedure
A. Officials will announce the initial pairing of pilots at the pre-flight briefing.

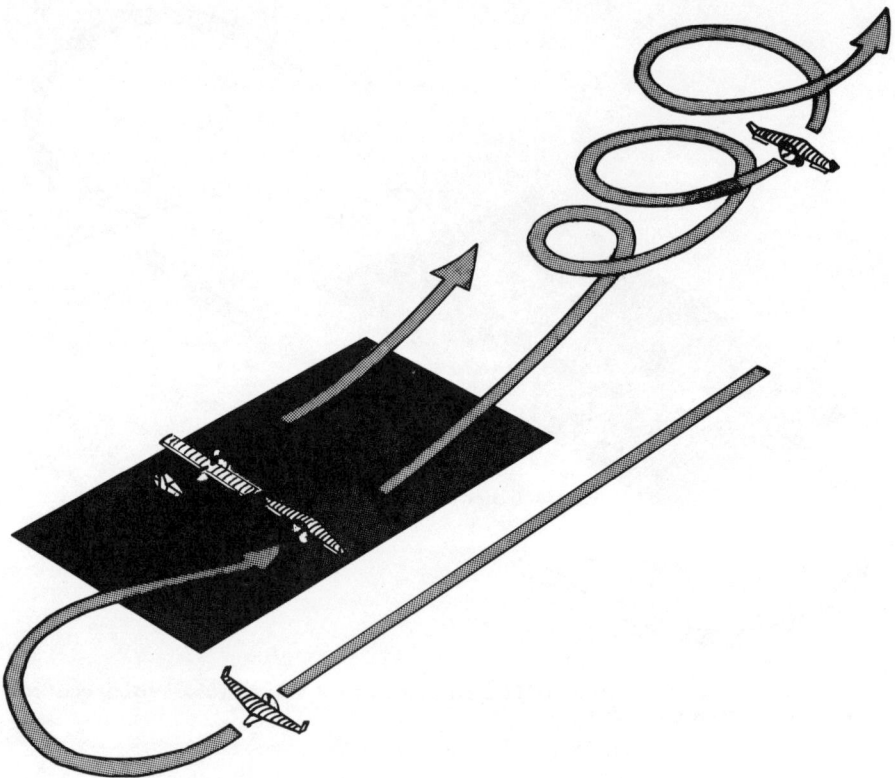

Fig. A-6. The "Accuracy Landing" competition is won by the pilot landing closest to a designated line.

B. Pilots will empty fuel tank and run-out engine under supervision before receiving measured fuel. A measured amount of fuel will be given.

C. Pilots will prime and start engine in readiness for their eventual flight.

D. Pilots position their aircraft for takeoff as directed.

E. Pilots will commence their takeoff run when signaled by the Flight Line Director and depart as briefed at the pre-flight meeting.

F. Pilots may land deadstick. In case of landing conflict, pilots will be informed by bullhorn or signal device and higher pilot may land outside designated area without penalty.

Scoring

A. The pilot of the pairing who is the last to land in the designated area winds the heat.

B. Failure to land in the designated landing area earns the pilot a loss for the heat unless otherwise authorized.

C. If both pilots fail to land in the designated landing area, the pilot landing closest to landing area wins.

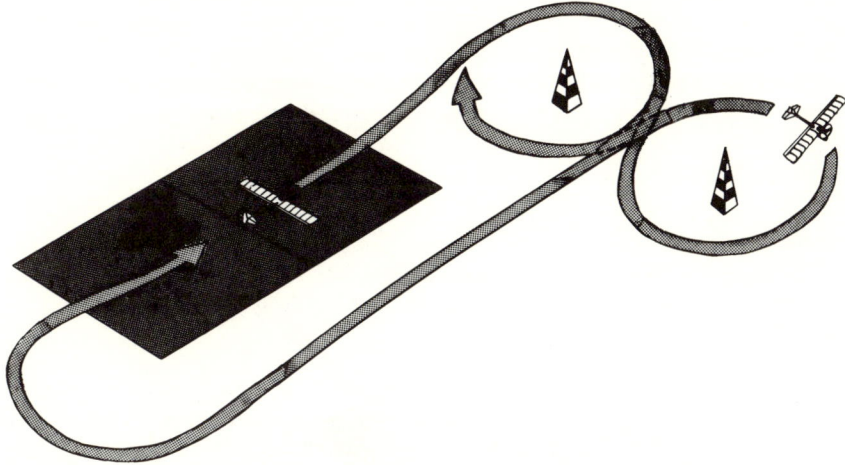

Fig. A-7. The pilot who flies a "Figure 8" around two pylons in the shortest time wins this event.

AOPA SANCTIONED TASK
FIGURE EIGHT MANEUVERS

FIGURE EIGHT MANEUVERS
This event tests the pilot's skill in safely performing a series of maneuvers.

General Description
A. Each flight consists of a takeoff, flight through the pylon course, and a landing. Pylons are placed 200 feet apart.

B. The object of the event is to safely perform, in the shortest time possible, a consecutive series of figure eights around pylons and land within a designated area. The Meet Director will designate the number of figure eights before each round.

Procedure
A. Officials will announce the pairing of pilots at the preflight briefing.

B. Pilots position their aircraft for takeoff as directed.

C. Pilot commences takeoff run when signaled by the Flight Line Director and flies the pattern as briefed at the preflight briefing.

D. Timing will begin when the pilot flies over the entrance/exit gate, and timing will end when the pilot completes the required number of consecutive figure eights and flies over the entrance/exit gate.

E. Pilot must pass to the outside of the pylon.

F. Pilot cutting a pylon must repeat that portion of the figure eight before continuing the maneuvers.

G. Pilot must cut power to idle before exiting the entrance/exit gate. Failure to cut power before exiting the gate results in a loss of the heat.

H. Pilot may re-apply power when abeam of a landing line.

I. Pilot must land within the designated area.

Scoring

Pilot with the shortest time within the maneuvering course and performing a landing within the designated landing area wins the heat.

AOPA SANCTIONED TIE-BREAKING TASKS
SHORT SHORT

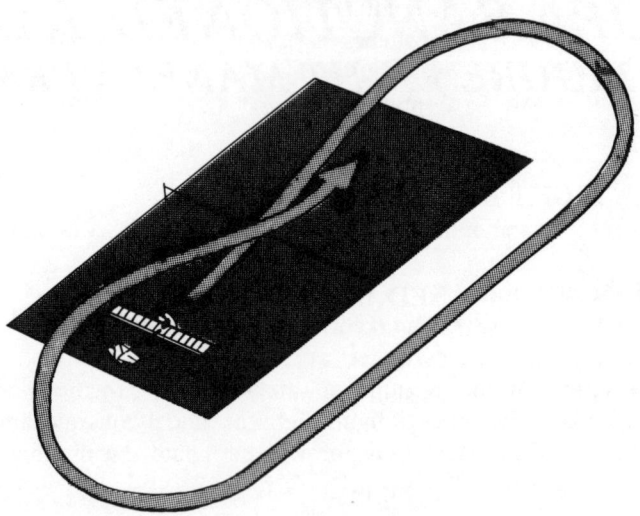

Fig. A-8. Called the "Short-Short", this is a tie-breaker event in which the winner is the pilot who can takeoff and land over a foot high barrier in the shortest distance.

THIS TASK TO BE USED ONLY AS A TIE BREAKER

SHORT SHORT

This event tests a pilot's knowledge of the performance of the aircraft and tests the pilot's landing skills.

General Description

A. The objective of the event is to take off, clear a barrier and land in the shortest distance possible.

B. Each flight consists of a takeoff, flight over the barrier, continuation of the flight around the pattern, and a landing beyond the barrier.

Procedure

A. Officials will announce the starting order of pilots at the preflight briefing.

B. The pilot places the aircraft on the runway at an estimated shortest distance possible to take off and clear a one-foot high barrier.

C. Pilot commences take off run when signaled by the Flight Line Director and flies the pattern as briefed at the preflight briefing.

D. Pilot must takeoff, clear the barrier, continue flight around the pattern, and land as closely as possible beyond the barrier.

Scoring

A. The shortest total distance used by an ultralight wins the heat. The distance shall be measured from the barrier to the beginning of the takeoff and from the barrier to the landing point.

B. A pilot whose aircraft touches or breaks the barrier (ribbon) receives a loss.

C. The ultralight must contain all parts as indicated by the manufacturer as constituting a complete machine.

GET DOWN

THIS TASK TO BE USED ONLY AS A TIE BREAKER

GET DOWN

This event tests the pilot's skill in safely maintaining low-level flight.

General Description

A. The objective of the event is to safely perform a flight that breaks a series of barriers without the aircraft touching the ground.

B. Each flight consists of a takeoff, flight through the low-level course, and a landing within a designated area.

C. The low-level course consists of five (5) barriers (ribbon) each no higher above the ground than eighteen (18) inches, at least twenty (20) feet wide, and separated from each other by at least twenty (20) feet distance. The course shall be arranged so that it may not be flown in a straight line.

Procedure

A. Officials announce the starting order of pilots at a preflight briefing.

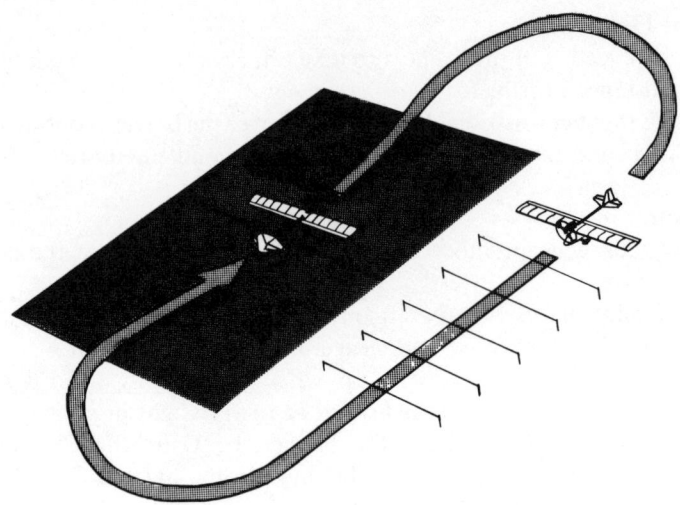

Fig. A-9. The "Get-Down" is another tie-breaker, which is won by the pilot who can break the most 18 inch high ribbons without touching the ground.

B. Pilots position their aircraft for takeoff as directed.

C. Pilot commences takeoff run when signaled by the Flight Line Director, flies the pattern as briefed at the preflight briefing, enters and exits the low-level course, and lands in the designated landing area.

Scoring

A. The pilot breaking the largest number of barriers without any part of the aircraft touching the ground after takeoff and prior to landing and who lands within the designated area wins the heat.

B. A pilot whose aircraft touches the ground, after takeoff and prior to landing in the designated landing area, receives a loss.

C. A pilot landing outside the designated landing area receives a loss.

A-11
The Powered Ultralight Manufacturers Association

The Powered Ultralight Manufacturers Association (PUMA) serves as the collective voice of powered ultralight flying. Founded in 1980, PUMA has become the largest association of ultralight manufacturers in the world. Today it represents about 90 percent of all ultralights manufactured in North America.

And PUMA's work is vital in ensuring the growth of ultralight flight as a recreational pursuit, free from unneeded government regulation.

The Federal Aviation Administration has placed the burden for developing airworthiness standards, pilot certification programs and vehicle registration programs squarely on the shoulders of the ultralight community. In the FAA's own words:

> The ultralight community is expected to take positive action to develop these programs in a timely manner and gain FAA approval for their implementation. Should this approach fail to meet FAA safety objectives, further regulatory action will be necessary. (Federal Register, Sept. 2, 1982)

That is the challenge. And PUMA is meeting the challenge. As the voice of the ultralight community, PUMA is advocating ultralight flight as a safe and enjoyable sport, monitored by a responsible ultralight community.

PUMA is working in many ways to promote ultralight flight and to aid the development of the ultralight vehicle industry, including:

- developing an airworthiness standard and vehicle certification program for ultralight vehicle manufacturers;
- keeping its members informed about developments affecting their industry and the sport; and
- representing the interests of its members at all levels of government.

PUMA will continue to be the voice of the ultralight industry in North America — working with government to keep ultralight flight a safe and enjoyable sport, monitored by and for ultralight enthusiasts.

Working with PUMA

Membership in the Powered Ultralight Manufacturers Association is open to all manufacturers of powered ultralights, and to suppliers of components and services.

PUMA members are on the ground floor of the booming ultralight industry, setting the course for the future of this sport. They're developing the standards against which the quality and performance of their craft can be measured — standards that will become increasingly important in consumer buying decisions. PUMA members are making sure their voices will be heard in the ultralight community.

From every perspective, PUMA is **the** organization for the ultralight industry that will determine the future of this fast-growing sport.

For more information about PUMA membership (without obligation) please write or call:

Roy W. Muth, President
Powered Ultralight Manufacturers Association
7535 Little River Turnpike, Suite 350
Annandale, Virginia 22003
Phone: 703-642-5859

A-12
Canadian Ultralight Pilot License Requirements

Ed's Note: As of April 1, the following rules governing ultralight aircraft student pilot permits, private pilot licenses and commercial pilot licenses went into effect in Canada.

STUDENT PILOT PERMIT

1.

2. An ultralight aeroplane student pilot permit certified for solo flight is required prior to commencing the solo flight training for issue of an ultralight pilot license.

3. An applicant for an ultralight aeroplane student pilot permit shall meet the following requirements:

(a) age

An applicant shall have reached his 14th birthday;

(b) medical fitness

(i) an applicant who meets the conditions specified on the Civil Aviation Medical Declaration (Form 26-0297 English or 26-0301 French) may sign that declaration,

(ii) an applicant who does not meet the conditions specified on the Civil Aviation Medical Declaration shall require a medical examination from a civil aviation medical examiner, or

(iii) an applicant shall have completed the medical examination requirements in accordance with the medical standards for civil flight crew licensing and be in possession of a valid License Validation Certificate;

(c) citizenship

An applicant must meet the conditions specified in subsection 2(a) of Chapter 1;

(d) knowledge

An applicant shall have passed a pre-solo examination including air regulations, air navigation orders, air traffic rules and procedures, information circulars, NOTAM, basic aerodynamics applicable to the type of ultralight being used for the training, meteorological phenomena as applicable and stall recognition and recovery procedures. This examination shall be administered by the holder of an ultralight aeroplane commercial pilot license and shall include questions specified in the appropriate Transport Canada Study and Reference Guide;

(e) experience and skill

An applicant shall have his pilot log book certified by the holder of an ultralight aeroplane commercial pilot license to the effect that the applicant has reached a satisfactory standard of experience and skill to complete solo flight.

Issue and certification of a student pilot permit.

(a) a student pilot permit shall not be issued unless evidence is provided that the applicant has met all requirements specified in section 3, the applicant has completed at least part A of an application for an ultralight aeroplane student pilot permit (Form 26-0297 English or 26-0301 French) and the holder of an ultralight aeroplane commercial pilot license, has completed parts B and C of that same form;

(b) following completion of the requirement specified in (a), the holder of an ultralight aeroplane commercial pilot license shall certify the student pilot permit for solo flight privileges by entering all required information on the student pilot permit and signing it. The applicant shall also sign the permit prior to solo flight.

5. Privileges.

The holder of a valid ultralight aeroplane student pilot permit may, for the purpose of his own flight training, act as pilot-in-command of any ultralight aeroplane provided that:

(a) he is under the supervision and direction of the holder of an ultralight aeroplane commercial pilot license;

(b) no passengers are carried and all flights during which he so acts are conducted
(i) within Canada,
(ii) under day VFR conditions, and
(iii) outside controlled airspace, unless otherwise authorized.

6. Permit validity.

(a) the normal validity period of an ultralight aeroplane student pilot permit shall be 60 months, calculated from the first day of the month following the date of medical declaration or medical examination; and

(b) an ultralight aeroplane student pilot permit shall only be valid for that category of aircraft. A student wishing to receive training on another category of aircraft shall meet all requirements for issue of a second student pilot permit valid for that other category.

7. Revalidation.

Revalidation of a student pilot permit is accomplished by meeting again the requirements specified in subsections 3(b) and 4(b).

PRIVATE PILOT LICENSE

1. age

An applicant shall have reached his 16th birthday.

2. Medical fitness

(a) an applicant who meets the conditions specified on the Civil Aviation Medical Declaration (Form 26-0297 English or 26-0301 French) may sign that declaration;

(b) an applicant who does not meet the conditions specified on the Civil Aviation Medical Declaration shall require a medical examination from a civil aviation medical examiner; or

(c) an applicant shall have completed the medical examination requirements in accordance with the medical standards for civil flight crew licensing and be in possession of a valid license validation certificate.

3. Knowledge

(a) except as otherwise provided in subsection 4(b), an applicant shall have demonstrated his knowledge by satisfactory completion of

(i) a course of ultralight aeroplane ground school which shall be in accordance with a training syllabus prepared by the ultralight aeroplane manufacturer or school and approved by the minister, and

(ii) a Transport Canada written examination on air regulations, air navigation orders, air traffic rules and procedures, information circulars and NOTAM based on the questions selected from the appropriate Transport Canada Study and Reference Guide;

(b) applicants holding valid pilot licenses in aeroplane and helicopter categories may be exempted from the examination specified in subsection (a) (ii).

4. Experience.

(a) all civil flight training, including dual instruction for this license shall be under the direction and supervision of the holder of a valid ultralight aeroplane commercial pilot license;

(b) an applicant shall have acquired in ultralight aeroplanes

(i) at least five hours flight time, including not less than one hour as sole occupant in ultralight aeroplanes, and

(ii) the flight time specified in (i) shall include not less than 25 takeoffs, full circuits and landings;

(c) an applicant who is the holder of or has held a valid pilot license of the aeroplane category within the preceding five years, may have the requirements specified in (b) reduced to not less than one hour as sole occupant and 10 takeoffs, full circuits and landings.

5. Recording of flight time.

Training flight time shall be recorded in a recognized pilot log book by the student. The entries shall be certified as correct by the holder of an ultralight aeroplane commercial pilot license. For flight time in other than club or school aircraft, the registered owner of the aircraft must certify the applicable entries in the pilot's log book.

6. Crediting of flight time.

The flight time acquired in ultralight aeroplanes shall not be credited toward any other aviation personnel license or endorsement thereto.

7. Skill

An applicant shall submit a letter from the holder of a valid Canadian ultralight aeroplane commercial pilot license stating that the applicant has demonstrated his ability to perform both normal and emergency manoeuvres appropriate to the ultralight aeroplane used for the training course and with a degree of competency appropriate to that of an ultralight private pilot.

8. Privileges.

The holder of a valid ultralight aeroplane private pilot license may, under day VFR conditions, act as pilot-in-command of any ultralight aeroplane provided no passengers are carried.

9. Revalidation.

(a) the normal medical validation period of the license is 60 months, calculated from the first day of the month following the date of medical declaration or medical examination. Revalidation of the license is accomplished by meeting the medical fitness requirements specified in section 3 and issue of a new license;

(b) to ensure continuing validity, application for renewal should be made during the 60 days immediately preceding the "valid to" date shown on the license.

COMMERCIAL PILOT LICENSE

1. An applicant shall have reached his 18th birthday.

2. Medical fitness.

(a) an applicant who requires in-flight instruction privileges shall have completed the medical examination requirements in accordance with the medical standards for civil flight crew licensing and be in possession of a valid license validation certificate category 1 or 3;

(b) an applicant who requires only ground monitored flight instruction privileges shall meet the requirements specified in (a) or one of the following:

(i) completion of a Civil Aviation Medical Declaration (Form 26-0297 English or 26-0301 French), or

(ii) if unable to sign the declaration, then part E may be completed by a Civil Aviation Medical Examiner.

3. Knowledge

(a) an applicant shall have successfully completed a course of instructional technique of not less than 10 hours, including

(i) the practical application of the basic principles of learning and techniques of instruction,

(ii) preparation and use of lesson plans,

(iii) flight preparatory instruction,

(iv) pre and post flight briefing procedures relative to air exercises and weather conditions, and

(v) normal and emergency manoeuvres;

(b) the course specified in (a) shall be presented in accordance with a syllabus of instructor training developed from section 1 of the Transport Canada Flight Instructor Guide, Aeroplanes. The course of training prepared by the ultralight aeroplane manufacturer or school shall be submitted to Transport Canada for approval;

(c) an applicant shall have demonstrated his knowledge by satisfactory completion of a written examination covering instructional techniques based on section I of the Transport Canada Flight Instructor Guide, Aeroplanes (Principles of Learning and Techniques of Instruction).

4. Experience.

(a) all civil flight training for this license shall be under the direction and supervision of the holder of an ultralight aeroplane commercial pilot license and presented in accordance with a Ground and Air Instruction Syllabus developed from section II of the Transport Canada Flight Instructor Guide, as applicable to the type of ultralight aeroplane being used for the training;

(b) an applicant shall have acquired in ultralight aeroplanes at least 20 hours total flight time, including not less than 10 hours as sole occupant of the aeroplane;

(c) an applicant who is the holder of or has held a valid pilot license of the aeroplane category within the preceding five years may have the requirements in (b) reduced to at least 10 hours, including not less than two hours as sole occupant.

5. Recording of flight time.

Flight time shall be recorded in a recognized pilot log book by the applicant. The entries shall be certified correct by the holder of an ultralight aeroplane commercial pilot license who provided the training. For flight time in other than club or school aircraft, the registered owner of the aircraft must certify the applicable entries in the pilot log book.

6. Crediting of flight time.

The flight time acquired in ultralight aeroplanes shall not be credited toward any other aviation personnel license or endorsement thereto.

7. Skill

An applicant shall submit a letter from the holder of a valid ultralight aeroplane commercial pilot license stating that the applicant has demonstrated his ability to perform both normal and emergency manoeuvres appropriate to the ultralight aeroplane used for the training course, and with a degree of competency appropriate to that of an ultralight aeroplane commercial pilot.

8. Privileges.

The holder of a valid ultralight aeroplane commercial pilot may

(a) act as pilot-in-command of any ultralight aeroplane under day VFR conditions provided no passengers are carried;

(b) give ground based flight instruction for any ultralight aeroplane provided he has at least five hours flight time in that type of ultralight aeroplane;

(c) give dual flight instruction in any ultralight aeroplane provided he holds a Medical Category 3 or higher License Validation Certificate and he has at least five hours flight time in the ultralight aeroplane;

(d) authorize ultralight aeroplane students for solo flight;

(e) certify that the holder of an ultralight student pilot permit has reached the competency for issue of an ultralight private pilot license; and

(f) certify that the holder of an ultralight private pilot license has reached the competency for issue of an ultralight commercial pilot license, except that certification for dual flight instruction privileges shall only be made by the holder of an ultralight commercial pilot license granted that privilege.

9. Revalidation.

The normal medical validation period of license is 60 months calculated from the first day of the month following the date of medical declaration or medical examination. Revalidation and issue of a new ultralight aeroplane commercial pilot license is accomplished by:

(a) meeting the medical fitness requirements specified in section 3; and

(b) submission of a letter of recommendation to Transport Canada from the holder of a valid ultralight aeroplane commercial pilot license, certifying that the applicant is familiar with current, instructional techniques and is competent to act as an ultralight aeroplane instructor.

A-13
Ultralight Organizations and Publications

ORGANIZATIONS

AOPA, Air Safety Foundation
7315 Wisconsin Avenue
Bethesda, MD 20814
(301) 951-3973

EAA, Ultralight Division
P.O. Box 229
Hales Corners, WI 53130
(414) 425-4860

Powered Ultralight Mfrs. Assn.
7535 Little River Turnpike
Suite 330
Annandale, VA 22003
(703) 642-5850

MOPAC
4206-38th Street
Edmonton, Alberta
Canada T6L 4K4

BOOK PUBLISHERS

Ultralight Publications, Inc.
P.O. Box 234
Hummelstown, PA 17036
1-800-441-7527 (only for credit card orders or our FREE Catalog)

Canadian Dist. for above
Acfield Aviation Supplies, Ltd.
7040 Torbram Rd. (14)
Mississauga, Ont. L4T 3Z4
(416) 677-4717

MAGAZINE PUBLISHERS

Air Progress Ultralights
7950 Deering Avenue
Canoga Park, CA 91304

Canadian Ultralight News
P.O. Box 563
Station "B"
Ottawa, Ont. K1P 5P7

EAA Ultralight Division
P.O. Box 229
Hales Corners, WI 53130

Flight Line Magazine/BMAA
11 School Hill
Wrecclesham, Farnam
Surrey, ENGLAND

Glider Rider
P.O. Box 6009
Chattanooga, TN 37401

Hang Gliding
P.O. Box 66306
Los Angeles, CA 90066

Sportsman Pilot
P.O. Box 485
Hales Corners, WI 53130

Ultralight Aircraft Magazine
16200 Ventura Blvd.
Suite 201
Encino, CA 91436

Ultralight Flyer Newspaper
P.O. Box 98786
Tacoma, WI 98499

Ultralight Pilot
P.O. Box 5800
Bethesda, MD 20814

Whole Air Magazine
P.O. Box 144
Lookout Mtn., TN 37350

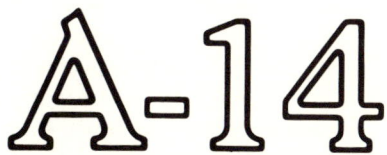

Insurance Exposures and Definitions for Ultralight Pilots and Organizations

Liability: Webster's definition says that the word means; "legally bound, as to make good any loss or damage that occurs in an action or transaction, to have responsibility, to be answerable." A liability insurance policy is a contract in which the insurance company agrees to **assume** the insured's liability either generally or in a specific area. Ordinarily, liability insurance covers Bodily Injury and Property Damage done to another party. Almost all liability insurance policies are written because the applicant wishes to protect his or her assets or a corporation's assets from suit. A liability insurance policy will also pay for the **defense** of the insured party. Many times suits are brought which are spurious or the prosecuted party is clearly not liable. Although suit may be dropped or dismissed by the court, an attorney to defend the insured must be paid.

Different risks call for different kinds of liability protection. For example, an ultralight manufacturer needs a different kind of protection than does an ultralight mechanic because they actually do different things. The manufacturer fabricates, and offers for sale an ultralight. The mechanic assembles, repairs and maintains that aircraft after it has been made and sold. The manufacturer needs products liability coverage and the mechanic should carry completed operations liability insurance.

The liability exposures facing the ultralight community are many and varied. Primary is the coverage carried by ultralight pilots for the liability arising out of the use of an ultralight. This coverage is basically the same as the liability insurance bearing on the operation of automobiles and standard aircraft. Bodily injury and/or property damage done while operating an ultralight by the insured would be covered by this policy. Some policies are restricted to use of a designated ultralight or to only the use of other ultralights.

Flying site liability is very different in nature. This exposure is primarily a premises risk; that is the policy will respond to injury and damage arising from the nature and maintenance of the property. An example of this might be a suit brought because of a fall at the site which resulted in an injury to a visitor. Almost always, if the property is rented or leased, the landowner is also named as an additional insured on the policy.

Some policies include an endorsement extending coverage for personal injury. Personal injury insurance covers areas like wrongful arrest, slander, libel, trespassing, and wrongful entry, etc. An organization might wish to carry this if they have a regular publication.

Another liability exposure some ultralight persons could have is teaching newcomers to the sport how to fly ultralights. This particular exposure calls for a professional type liability coverage such as that carried by physicians, attorneys, insurance agents, and other professionals. The coverage applies to the acts or omissions of the instructor both during instruction or after the fact. An example might be a suit arising from an accident where the plaintiff sues an instructor for not having told him what to do in the given situation which allegedly caused the accident.

For ultralight dealers, several other commercial type liability exposures may be present. Among these could be: premises liability for a store location separate from the flying site, completed operations liability if you do assembly or repair work for others, general liability to cover your sales activity, air show displays, etc., hanger keepers type liability to cover other peoples property (ultralights) that are left in your care, custody and control, and other exposures as they might appear for a dealer. A dealership is a business every bit as much as an auto garage, store, etc.

There are several other liability exposures which may have to be considered from time to time by dealers or site operators. Contractual liability which protects the policy holder from liability described in a contract rather than that imposed by law. Owners protective liability insures against loss due to the operations of sub contractors. Umbrella liability coverage picks up where the basic liability coverages end and generally are written in amounts of at least $1,000,000 and higher. It sometimes is called excess liability coverage because it pays after the primary insurance has been exhausted.

Occasionally, an unusual risk comes up which requires special attention. An excellent example of this would be a public fly-in or air show. The exposure here far exceeds the provisions of coverage to say, a flying site policy. Ordinarily, a flying site may have up to one hundred people at the site either flying or observing. At an air show type function, suddenly the number of people there is drastically increased. There is additional auto parking, perhaps food service, bleachers, and sometimes an admission charge. Flying events take place at the site which are not the kinds of events that occur normally on a daily basis. In short, we have a very different exposure from the ordinary.

This exposure can be very simply handled by endorsing the existing site policy with a special events endorsement. The endorsement usually involves an additional charge, but it will cover all the different risks to which the site operator may find himself exposed. The landowner and air show/fly-in sponsor can also be covered if desired.

Waivers of liability are a subject of concern to anyone involved in any kind of aviation activity, not just ultralighting. At one time, a signed and witnessed disclaimer was sufficient to defer any liability. The courts have long since dispelled the use of waivers to avoid liability. However, this is not to say that they should not be used. The use of signed waivers is encouraged for the following reasons, among others; first, the waiver shows in writing that the signer knows and acknowledges that the activity is dangerous. It also shows that he agrees to assume responsibility for his actions. In case of an accident, it often dissuades the injured person from beginning a legal action.

In ultralighting a waiver might include some language stating that the signer acknowledges that an ultralight is not an aircraft and they have distinctly different flight characteristics. Also, it might say that the pilots in flight are subject to natural forces beyond their control and that accidents occurring as a result of these forces are not the responsibility of the issuer of the waiver.

There are other insurance exposures such as physical damage coverage for the ultralight itself and fire type coverages for dealers or clubs, buildings, contents or equipment in storage. These exposures may be covered for specified perils such as fire and theft or for all risks. All risk coverage traditionally covers property for any occurrence except a few published exceptions, such as wars and nuclear explosions. Normally, deductibles are used for property coverages to help control the risk and to make the coverage affordable. Coverage for an ultralight's "hull" is a physical damage coverage.

Coverage for other property such as buildings, equipment and stock are written as fire type insurance. All lenders require physical damage insurance be carried on anything on which they have a lien.

All in all, insurance has become an integral part of all our personal, association, and business activities. It is necessary for the real growth and credibility of the ultralight sport and movement. Carrying liability insurance is an expression of responsibility and prudence and is regarded as such by the courts and the public.

Ed. Note: Home owners policies do not cover ultralight flying. You wouldn't think of driving your car without proper insurance. Likewise, having proper ultralight coverage is absolutely essential. You wouldn't want your favorite recreation to ruin your life.

For more information on insurance, contact: Lightwing Insurance, P.O. Box 16, Westerville, OH 43081 (614) 882-5135.

The Ultralight Library
Practical Aviation Books From Ultralight Publications

1
ULTRALIGHT AIRCRAFT - The Basic Handbook of Ultralight Aviation (Revised 2nd Edition) by Michael A. Markowski. This is the best selling (over 35,000 sold), definitive word on ultralight flying and aircraft. Divided into four sections, it covers: **Ultralight Aircraft Described,** including - Specifications to over 50 Aircraft ● Performance ● Handling ● Drawings ● Pilot Reports ● **The Basic Ultralight Flight Manual** describes the specialties of ultralight flying - Principles ● Stability and Trim ● Low Speed Flight Control Techniques ● Stalls ● Spins ● Landing ● Traffic Pattern ● Slideslipping ● Crosswinds ● Crabbing ● Instruments ● Density Altitude ● The Koch Chart ● Wind Chill Factors ● Navigation ● Flight Planning ● Winds and Weather ● **Ultralight Propulsion** includes - Engine Operation ● Trouble-Shooting ● Engine Reviews ● Propellers ● **Appendicies** cover - Test and Study Guide ● FARs ● Manufacturers and Dealers Lists ● Plus much, much more. Ultralight Aircraft is highly recommended and endorsed by industry leaders. "Every aspect of ultralight aviation is covered by a professional who knows!" 320 pgs., 220 ill., 6x9 in.
Order No. 1.................Hardbound $22.95............Paper $15.95

2
ULTRALIGHT AIRMANSHIP - How To Master The Air In An Ultralight, by Jack Lambie. What can you do after you've learned the basics of ultralight flight? What's next? This exciting new book spells it out in clear, concise language — how you should fly to make use of, avoid, and operate in various atmospheric conditions, with specific advice and flight descriptions by experienced flyers. Large weather systems and circulation patterns, as well as the intricacies of micrometeorology (airflows around valley passes, mountains, behind trees, buildings and other obstructions) are described in detail. Learn how to handle turbulence and fly practically anytime you want to — you don't have to limit yourself to just mornings and evenings. If you know what's in this book! ULTRALIGHT AIRMANSHIP is your ticket to total mastery of the air — your "roadmap-to-the-sky." 144 pgs., 110 ill., 6x9 in.
Order No. 2.................Hardbound $18.95............Paper $10.95

3
ULTRALIGHT FLIGHT - The Pilot's Handbook of Ultralight Knowledge, by Michael A. Markowski. Covers the new world of ultralight aerodynamics, stability and control, design and performance, plus a fascinating history. If you want to be the most complete pilot you can be, you must know more than just how-to fly! The competent pilot needs to know the hows and whys of flight, as well as how ultralights evolved. His natural curiosity and quest for self-improvement, safety and his ultimate enjoyment of flying drives him to learn as much as he can about his machine's interaction with the air. Like none before, this book presents a vitally important subject in an easily understood manner. Lift, drag, thrust and weight, as well as stability, control and design are described in detail - without math! Learn why your ultralight flies the way it does. Be able to predict how any design will handle. Discover the intriguing early days of ultralights to gain an appreciation of why they are built the way they are. ULTRALIGHT FLIGHT is your key to unlocking the mysteries of flight - your guide to "intelligence-in-the-sky," written as only Markowski can! 224 pgs., 110 ill., 6x9 in.
Order No. 3.................Hardbound $20.95............Paper $13.95

4 **ULTRALIGHT PROPULSION - The Basic Handbook of Ultralight Engines, Drives and Propellers, by Glenn Brinks.** If you expect to fly with utmost confidence, you must know your power system — its characteristics, and idiosyncracies — inside out. The two-cycle ultralight engine is a stranger to most pilots, but it must be thoroughly understood before flying can be done with any degree of safety. ULTRALIGHT PROPULSION describes the incredibly important details of power — how it is produced and transmitted into the air to provide performance — from a pilot's point of view. The book describes ignition systems, carburetors, starters and starting, spark plugs and how-to read them, exhaust systems, break-in and trouble-shooting, teardown, inspection and reassembly, modification, accessories and controls. The various drive methods are reviewed — belts, gears and direct. The practical aspects of propeller operation, care, balancing, tracking and safety are presented. ULTRALIGHT PROPULSION tells you what power is all about — from a drop of gasoline to a rate of climb. It tells you how to get maximum performance, reliability and life from your power system. It's your most important *"tool"* for a good running engine. 224 pgs., 110 ill., 6x9 in.
Order No. 4 .... Hardbound $20.95 .... Paper $13.95

THE ULTRALIGHT LOG BOOKS

Designed with the ultralight pilot in mind, each has space for over 500 flight entries. Each is bound in a beautiful, colorful, heavy leatherette cover for long wearing durability. Fits in hip pocket! These log books are the only way for you to keep track of your valuable flight experience, aircraft usage, and engine operation. Know your proficiency level as well as your ultralight's condition.
THE ULTRALIGHT PILOT FLIGHT LOG, includes: Pre-flight inspection procedures, load factor and stall speed vs bank angle chart, ultralight traffic pattern, density altitude/Koch chart, landing factors and field perspectives, wind chill chart, and the flying-speed-of-an-ultralight.
Order No. P-1 .... $3.95
THE ULTRALIGHT ENGINE LOG, includes: Trouble-shooting guide.
Order No. E-1 .... $3.95
THE ULTRALIGHT AIRCRAFT LOG
Order No. A-1 .... $3.95
ULTRALIGHT CROSS-COUNTRY PLANNING AND LOG SHEETS (50 forms)
Order No. C-1 .... $1.95
BUMPER STICKER: FLY AN ULTRALIGHT AIRCRAFT - DISCOVER A NEW DIMENSION IN FLIGHT (3"x12": Yellow & Blue on White)
Order No. BS-1 .... $1.59
T-SHIRT (Same wording as bumper sticker, S, M, L, XL - four color)
Order No. TS-1 .... $10.95

- -

TO ORDER: Put your name and address on a sheet of paper. List the book title and order number, and include cash, check or money order to cover book cost plus $2.95 for postage and handling of total order. A *FREE* catalog is included with each order.
Send to: Ultralight Publications, P.O. Box 234, Dept. UP, Hummelstown, PA 17036 USA.